Foams - Emerging Technologies

Edited by Huijin Xu,
Chen Yang and Dengwei Jing

Published in London, United Kingdom

IntechOpen

Supporting open minds since 2005

Foams - Emerging Technologies
http://dx.doi.org/10.5772/intechopen.77942
Edited by Huijin Xu, Chen Yang and Dengwei Jing

Contributors
Anastasia Sychova, Larisa Svatovskaya, Maxim Sychov, José Miguel Molina Jordá, Lucila Paola Maiorano Lauría, Nicolai B. Jordanov, Esmat Hamzawy, Alexander Karamanov, Dragomir Tatchev, Ahmed Farid Ibrahim, Hisham A Nasr-El-Din, Huijin Xu, Chen Yang, Akira Nakayama, Shyam Sunder Sharma, Marco Caniato, Jan Kašpar, Giada Kyaw Oo D'Amore

First published in London, United Kingdom, 2020 by IntechOpen
IntechOpen is the global imprint of INTECHOPEN LIMITED, registered in England and Wales, registration number: 11086078, 5 Princes Gate Court, London, SW7 2QJ, United Kingdom
Printed in Croatia

British Library Cataloguing-in-Publication Data
A catalogue record for this book is available from the British Library

Additional hard and PDF copies can be obtained from orders@intechopen.com

Foams - Emerging Technologies
Edited by Huijin Xu, Chen Yang and Dengwei Jing
p. cm.
Print ISBN 978-1-78985-989-8
Online ISBN 978-1-78985-990-4
eBook (PDF) ISBN 978-1-83968-107-3

For EU product safety concerns:
IN TECH d.o.o., Prolaz Marije Krucifikse Kozulić 3, 51000 Rijeka, Croatia, info@intechopen.com or visit our website at intechopen.com.

Meet the editors

Dr. Huijin Xu is an associate professor at Shanghai Jiao Tong University. He received a bachelor's degree, master's degree, and PhD from Xi'an Jiaotong University, Xi'an, China. His research focuses on flow/heat/mass transport in porous media, thermal energy storage, thermochemical processes, and more. He has a good record of publications in the areas of thermal storage, porous media, and heat/mass transfer. He also has a good record in educating undergraduates and postgraduates with some approved achievements. Dr. Xu is currently an associate editor for *Special Topics & Reviews in Porous Media* and an editorial board member for the *Journal of Porous Media*.

Dr. Chen Yang is a professor at the College of Chemical Engineering, Fuzhou University, China. He received his PhD in 2012 from the University of Shizuoka, Japan. From 2012 to 2014, he worked as a postdoc at the National Polytechnic Institute of Toulouse (INPT), France. He joined Fuzhou University as an associate professor in 2014 and was promoted to full professor in 2016. His research activity covers various areas, such as transport theory in porous media and chemical process intensification. He has published more than forty papers in journals such as *AIChE Journal, Chemical Engineering Science*, and others. Due to the achievement of catalytic packing optimization, Dr. Yang received a number of awards including first prize for S&T Progress Award (2016, China Petroleum and Chemical Industry Association; 2018, People's Government of Fujian Province).

Dr. Dengwei Jing is a full-time professor at the International Research Center for Renewable Energy and the State Laboratory of Multiphase Flow in Power Engineering, Xi'an Jiaotong University, China. He obtained his PhD in Thermal Engineering from the same university in 2007. His field of expertise is renewable energy conversion and enhanced heat and mass transfer. Currently, he leads a large project funded by the National Key Research Program of China, and has obtained China National Funds for Excellent Young Scientists. He was awarded the Newton Advanced Fellowship from Royal Society, UK. Dr. Jing has published more than 130 peer-reviewed journal papers in *Nature Communications, Applied Physics Letters, AIChE Journal*, and others. He has been a plenary/keynote speaker at many renowned international conferences. He is now the associate editor for *Journal of Applied and Computational Mechanics* and is an editorial board member for many other journals, including *Thermal Science and Engineering Progress, Heliyon, Fluids, and Fluid Mechanics Research Journal.*

Contents

Preface

Within a few hundred years of the Industrial Revolution, humans discovered or fabricated thousands of foam fluids and materials that have been utilized in petroleum extraction, chemical engineering, the textile and architecture industries, and more. In addition to synthesized foams, there are considerable foams found in nature that are very stable evolutionary and bionic structures. Foam structures have been widely studied and applied in many subjects, and scholars in the field have played an indelible role in developing foam-related theories that are being used to figure out how to make better use of foams today.

Foams are ubiquitous in human life. For example, foams are found in sodas and sponges in liquid form and solid form, respectively. Various foams have distinctive properties that can be used to develop special usages in engineering applications. This book reviews, researches, and summarizes the knowledge and experience of foam fluids and porous foams in industry. Compared with simpler fluids, foam fluids are more complicated to describe by conventional rules because of their phase discontinuity. Considering that foam fluids can be applied to displace oil or gases underground, research on foam fluids is of great significance. Porous foams consist of solid metal and fluid-filled pores that take up a certain portion of the entire volume. This structural feature makes porous foams mechanically stable with lighter mass. In addition, porous foams have terrific performance when employed in energy absorption and heat exchange. Nevertheless, the pores existing in porous foams add the difficulties of modeling and calculating the fluid flow. Fortunately, this book addresses this issue and provides possible solutions.

Chapter 1 reviews recent developments in the manufacture and characterization of multiphase foams developed by incorporating new phases into open-pore foam materials. The new incorporated phases can significantly alter the macro-/microstructure of the starting materials or modify the pore surfaces to achieve new functionalities, which exhibits a great potential for use in electronics, medicine, or catalysis.

Chapter 2 discusses state-of-the-art acoustic and thermal models and their application to cellular foam materials. Five different forecasting methods including traditional analytical, a modified analytical with a new proposed equation, and inverse procedures were employed to determine the Johnson–Champoux–Allard (JCA) parameters related to the sound-absorbing properties of foam materials. Numerical results indicate that the inverse procedure, using the thermal characteristic length derived from the scanning electron microscope (SEM) micrographs as the imposed parameter, well agrees with the modified analytical model.

Chapter 3 presents a turbulent heat transfer analysis of silicon carbide ceramic foam as a solar volumetric receiver. Both the Rosseland approximation and the P1 model are applied to consider the radiative heat transfer through the solar receiver. In light of the derived analytical solutions, it is found that the corresponding fluid and solid temperature variations generated under the Rosseland approximation agree fairly well with those based on the P1 model. Furthermore, optimal pore diameter that exists for achieving the maximum receiver efficiency under the equal pumping power is obtained, which provides effective guidance for a novel volumetric solar receiver design of silicon carbide ceramic foam.

Chapter 4 shows how to obtain sintered self-glazed glass-ceramics and/or glass ceramic foams with improved properties and differences in structure by means of a double-stage heat treatment. The sintering of the samples is studied by optical dilatometry and the foaming process by hot-stage microscopy, while the structure of the final materials is revealed by 3D computed tomography and SEM. The phase composition of the glass-ceramic foams is analyzed by XRD. Due to excellent material characteristics, the synthesized inorganic glassy-crystalline foam materials can be used as low-cost thermal insulating, soundproofing, and fire-resisting low-weight materials.

Chapter 5 uses casting techniques to design a new foam product and develop a new method of production to overcome shortcomings. The knowledge gained from this work is valuable for planning future actions for further improving aluminum metallic foam.

Chapter 6 proposes a novel approach to increase stability by introducing nano-sized additives SiO_2 and $Fe(OH)_3$ sols to improve the quality of construction foam on a protein basis for non-autoclaved foam concrete. Various stabilization mechanisms related to different energies of chemical bonds formed between the molecules of the foaming agent and the injected sols have been found. The stabilization of the construction foam leads to the possibility of using foam concrete hardening accelerators without destroying its structure, which is beneficial to obtain increased compressive and bending tensile strength and reduce thermal conductivity and shrinkage in drying.

Chapter 7 highlights the different CO_2-foam generation mechanisms and the deterioration effect of crude oil on CO_2-foam stability, so as to improve sweep efficiency in enhanced oil recovery (EOR) applications over that of polymers. Several nanoparticles such as aluminum oxide, copper oxide, and silicon dioxide are considered. The results indicate that silicone dioxide with a modified surface is more effective in foam stability applications.

In each of these chapters, we spared no efforts to provide sufficient and logical information for readers, and tried our best to establish the foundation of relevant research topics. Experts in the field reviewed each chapter of this book. As the author of this book, I am looking forward to receiving feedback and constructive comments from readers and experts.

Huijin Xu
China-UK Low Carbon College,
Shanghai Jiao Tong University,
Shanghai, China

Chen Yang
College of Chemical Engineering,
Fuzhou University,
China

Dengwei Jing
International Research Center for Renewable Energy,
State Key Lab of Multiphase Flow in Power Engineering,
Xi'an Jiaotong University,
China

Chapter 1

Open-Pore Foams Modified by Incorporation of New Phases: Multiphase Foams for Thermal, Catalytic and Medical Emerging Applications

Lucila Paola Maiorano Lauría and José Miguel Molina Jordá

Abstract

Recently, open-pore foam materials have acquired great interest in several technological sectors due to their excellent properties of low density, great specific surface area, adjustable thermal conductivity, and high-energy absorption. The replication method has proved to be one of the most widely used techniques for their manufacture, allowing a perfect control of the pores' characteristics from which the main properties of the foams derive. However, these properties have limited the use of these materials in ultimate applications of the most demanding emerging technologies. This chapter reviews recent developments of open-pore foams that have been modified by the incorporation of new phases in order to enhance their properties. The inclusion of new phases taking part of the microstructure or modifying the pore surfaces allows these materials to be considered promising for the most modern applications including, among others, thermal dissipation, catalytic supports, and medical implantology.

Keywords: open-pore foam, multiphase foam, inclusions, thermal management, catalysis, medical implantology

1. Introduction

Foam materials were originally conceived by clear inspiration in some natural porous materials, such as wood, bamboo canes, or bones, as they present a very attractive combination of properties such as excellent mechanical strength blended with low density [1]. Motivated by the versatility of those natural porous materials, human ingenuity succeeded in the design of new foam materials, their most suitable manufacturing processes, and their use in technologically demanding applications. In recent years, foam materials have reached a high level of maturity in their manufacture, development, applications, and integration into complex systems to fulfill specific applications.

Foam materials can be classified depending on their nature, pore interconnectivity, morphology of their cellular structure, or other variables that allow their differences to be outlined. A widespread classification divides foams into open-pore and closed-pore, depending on whether their porous cellular structures are interconnected or not, respectively. When more than half of the cells are open, the materials are considered open-pore foams. Closed-pore foams were proven useful in thermal insulation and structural applications (load-bearing components, energy absorbers, etc.) as well as in biomedical implants. Open-pore foams have cells that are not completely closed so a fluid can pass through the material. While open-pore foams are structurally less interesting, their open-pore space expands their utility to functional applications such as particle filters, bacteriological filters, active heat dissipation units, etc.

The fields of application of open-pore foam materials depend on their porous architectures and the nature of their solid phases. Open-pore ceramic foams have traditionally been used as thermal insulators, bio-scaffolds in tissue engineering, catalytic supports, and materials for sound and impact absorption, among others [2]. Recently, their use has been extended to catalytic applications given their suitability to be catalyst supports in gaseous or liquid phase reactions, since the presence of interconnected pores allows the passage of fluids and, therefore, can be used in continuous reactors. Polymer foams show excellent properties which make them suitable for many applications such as construction, cushioning and insulation, or sound dampening [3]. Open-pore metal foams also share some of these applications, but they deserve a special attention. Their outstanding mechanical and thermoelectrical conductive properties allow these materials being considered excellent candidates for a wide variety of applications depending on their porous structure, as it can be seen in **Figure 1**. Their high surface area per unit volume, low density, and great heat transfer capacity make them suitable for thermal management (heat exchangers and heat sinks), electrode materials, catalyst carriers, and biomedical engineering as biocompatible and biodegradable scaffolds [4]. When used in medical implantology, the interconnected structure provides a transition space between the bone and the biomaterial structural support, which allow the in-growth of bone tissue and vascularization [5]. Other properties such as high strength and toughness, great sound-absorbing capacity, and high impact energy

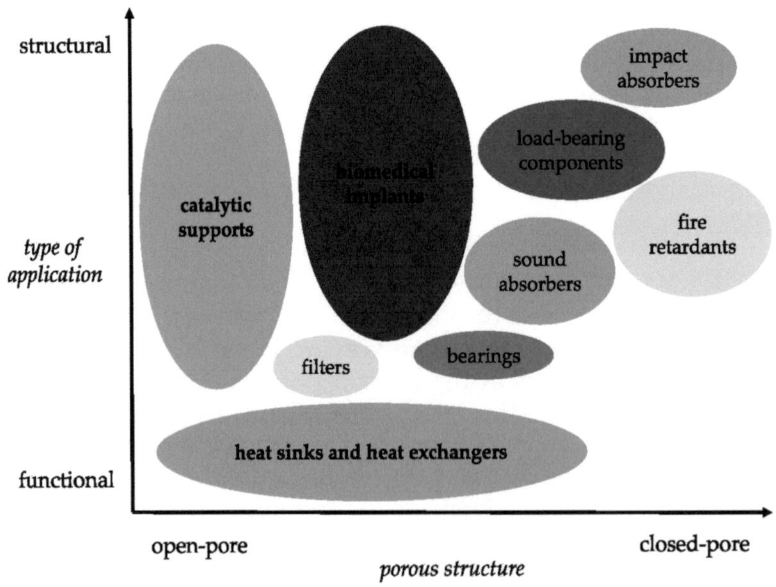

Figure 1.
Applications of metal foams according to porous structure. Partially reproduced from [1].

absorption make them interesting materials for structural applications in the aerospace, automotive, or marine industry.

Despite all their great attributes, traditional foams are often inappropriate to meet the requirements of the most advanced technological challenges; hence their designs have recently been reformulated by the incorporation of new functional phases. In this work, the authors focus the attention on this last type of open-pore foams, in which different components/phases have been incorporated to generate multiphase materials with a great potential of use in applications of different sectors such as electronics, medicine, or catalysis.

2. Current needs to incorporate new phases into open-pore foams

The development of emerging technologies such as new electronic devices in electronics, aeronautics, and aerospace; advances in the chemical industry; and the still incipient stage of biomedical engineering is concomitant with the accelerated progress of research into new materials. Some of the most demanding applications require further developments of open-pore foam materials and are discussed here below.

Thermal management has become a critical issue that often slows down or even hinders the progress of evolving power electronic technologies as a result of increasing power densities and decreasing transistor dimensions [4–8]. A successful strategy for efficient heat removal in electronic systems, called active thermal management, consists in forcing the direct transfer of heat from hot spots to some carrier fluids through a conduction-convection mechanism by means of using high thermally conductive open-pore foams. Research into new materials for these applications was focused mainly on metal and carbon/graphite foams as they exhibit interesting physical properties, such as low density and high specific area per unit volume, as well as decent thermal conductivity [9–12]. Many authors have focused on the investigation of forced convection parameters, such as heat transfer coefficient and pressure drop for different metal foams. Although these materials exhibit interesting characteristics, their properties of relatively high heat transfer coefficient and low pressure drop are still insufficient for its use in final applications of the most demanding emerging electronic technologies. Recent developments have modified open-pore foams by the incorporation of new phases either into the solid or into the cavities of the porous structure. These materials show considerable improvements in their thermal properties [8, 13, 14].

Open-pore foams used in catalytic applications must meet two requirements: a high specific surface allowing high dispersion of the catalytically active phase and not too small pore sizes to prevent a high pressure drop of the fluid passing through it. Although the open-pore foams used so far in catalysis have roughly met these characteristics, the new demands for better catalytic performance require materials with new structural pore designs and improved properties. To this end, porous materials must provide (i) the highest possible thermal conductivity to improve heat transport from or to the outside of the catalytic reactor (easily achievable when the nature of the solid is metallic) and (ii) the possibility to break the laminar flow in order to enhance the interaction between the fluid and the catalyst. The first requirement can be achieved by incorporating thermal inclusions into the solid phase of the foam and/or by a crystalline modification of the solid phase assisted by the catalytic action of the new present phases. The second requirement can be achieved by incorporating new phases into the porous cavities.

In addition to the mentioned applications, open-pore metal foams are recently being the subjects of intense study in medical implantology. These materials are not only intended to fulfill a structural purpose in a body system but also to cover functional applications. It was recently proposed to incorporate guest phases in the porous cavities of open-pore foams charged with pharmacological substances, with the aim to set a drug delivery system to avoid postsurgical infections [15, 16].

By way of the commented examples, the authors intend to highlight that the inclusion of new phases into open-pore foams opens up a range of new properties in foam materials and seems to be a suitable way to overcome the requirements of modern applications such as some of those commented for thermal management, catalytic chemistry, and medical implantology. In addition, some research works focus on the incorporation of new phases into foams to enhance mechanical properties as in all the mentioned applications, better mechanical performances are also soaked.

3. Manufacture

3.1 Manufacturing techniques of open-pore foams

Manufacturing techniques of open-pore foams can be classified into four groups attending to the state of the precursor material: liquid, solid, vapor, and ions [1].

Liquid state processing: the precursor material is in liquid state. The most important processing techniques are:

a. Investment casting with polymer foams

b. Casting around space holder materials/infiltration of martyr preforms

Solid state processing: the precursor material is in solid state. The following techniques are the most important ones:

a. Partial sintering of powders and fibers

b. Foaming of slurries

c. Pressurization and sintering of powders in martyr preform

d. Sintering of hollow spheres

e. Sintering of powders and binders

f. Reaction sintering of multicomponent systems

Vapor state processing:

g. Vapor deposition onto polymeric foams

Ionic solution state processing:

h. Electrodeposition onto polymeric foams

Despite the wide range of fabrication methods that these four groups generate, there are actually only two different strategies for generating porosity [17]:

- Self-formation: porosity is formed through a process of evolution according to the physical principles. Self-formation includes the d method.
- Predesign: the structure is created with the use of molds that determine the porous cavities. By means of this strategy, closed-pore (or not interconnected) and open-pore (or interconnected) foams can be manufactured, depending on whether the mold forms part of the final material or is removed, respectively. Predesign includes a, b, c, e, f, g, h, i, and j methods.

Among the manufacturing techniques, the infiltration of martyr preforms, also known as the replication (predesign) method, allows the best control over the material. This method was traditionally used to produce open-pore metal foams and recently adapted to produce carbon/graphite foams [18]. The replication method consists of the infiltration with molten metal or any other liquid precursors of a porous template preform that is later removed by dissolution or controlled reaction to leave a foam material with a porous structure that replicates the original preform. This method allows perfect control of size, shape, and size distribution of pores. Depending on the matrix material and the desired final porous architecture, different raw materials have been used as templates. Nevertheless, the most widespread martyr material is sodium chloride in particulate form, which can be conveniently packaged and infiltrated with liquid metals at temperatures below its melting point (801°C) and then removed by dissolution in aqueous solutions [13].

3.2 Manufacturing techniques of multiphase open-pore foams

The multiphase open-pore foam materials developed so far are still scarce and can be manufactured by various methods, which are reviewed in **Table 1** and later in the chapter.

3.2.1 Composite foams/foams with guest phases with preload of new phases in the preform

Loading of new phases is achieved by one of the following two strategies: (i) loading particles (inclusions) are packed together with larger martyr particles forming a porous bimodal preform, or (ii) loading particles are covered by a martyr material and packed forming a porous monomodal preform. Preforms are infiltrated and the martyr material is leached away. As a result, composite foams or foams with guest phases are obtained. They show homogeneous dispersion of new phases in a continuous matrix. **Figure 2** represents the aforementioned material structures. In particular cases, loading powders are packed combined with larger martyr particles and sintered. The martyr particles are later removed to obtain an interconnected porous structure.

3.2.2 Composite foams with preload of new phases in the liquid precursor

Loading of new phases is achieved by the dispersion of particles into the liquid precursor. The preform is leached away after its infiltration with the liquid precursor, and the final material shows a homogeneous dispersion of new phases in a continuous matrix. Material consolidation can also be obtained (instead of by infiltration) by electrochemical (co)-deposition of a metal and/or the new phases on a leachable preform or a preexistent porous material (the liquid precursor is an electrolyte that contains metal ions and dispersed particles of the new phases) (**Figure 3**).

Method	Inclusion into monolithic materials		Combination of monolithic materials		
Material type	**Composite foams/foams with guest phases**	**Composite foams**	**Finned foams**	**Monolithic finned foams**	**Composite finned foams**
New phase loading	Preload in preform	Preload in liquid precursor	No preload	No preload	Preload in preform
Distribution of new phases	Homogeneous dispersion in matrix	Homogeneous dispersion in matrix and/or pore surface	Homogeneous dispersion in one component; layered distribution of components	Homogeneous dispersion in one component; layered distribution of components	Homogeneous dispersion in one component; layered distribution of components
Assembly	Combination of packed/self-standing preforms with liquid precursor or powders	Combination of packed/self-standing preforms with liquid precursor/electrodeposition	Physical or glue joining of preexistent monolithic materials	Casting of liquid precursor in a mold	Combination of packed/self-standing preforms with liquid precursor
Matrix continuity	Continuous	Continuous	Noncontinuous	Continuous	Continuous
References	[8, 13, 15, 16, 19]	[8, 14, 20–29]	[30–32]	[33]	[13]

Table 1.
Methods currently developed to manufacture multiphase open-pore foams.

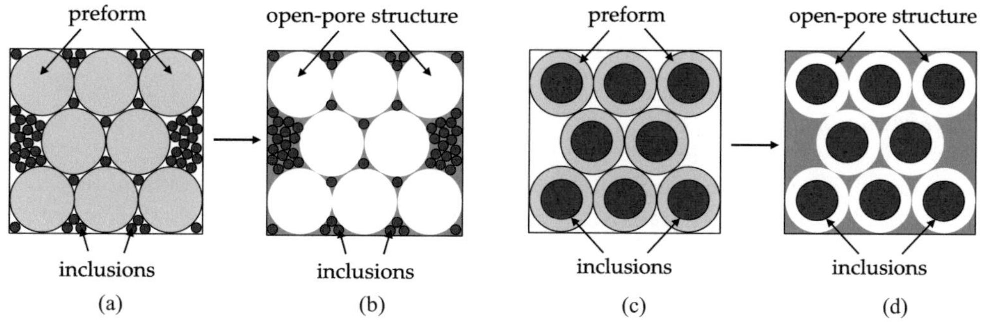

Figure 2.
Schematic drawings showing preform compositions (a and c) and the structures of the final materials (b and d) for composite foams (a and b) and foams with guest phases (c and d) obtained by preload of new phases in the preform.

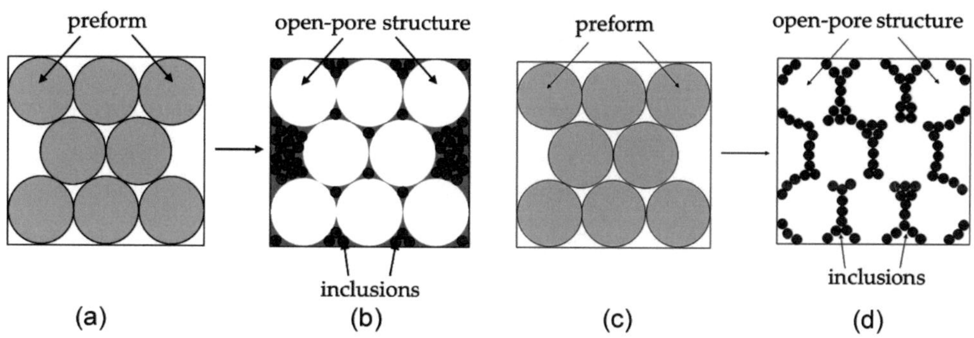

Figure 3.
Schematic drawings showing preform compositions (a and c) and the structures of the final materials (b and d) for composite foams obtained by preload of new phases in the liquid precursor processed by infiltration (a and b) or by electrochemical (co)-deposition (c and d).

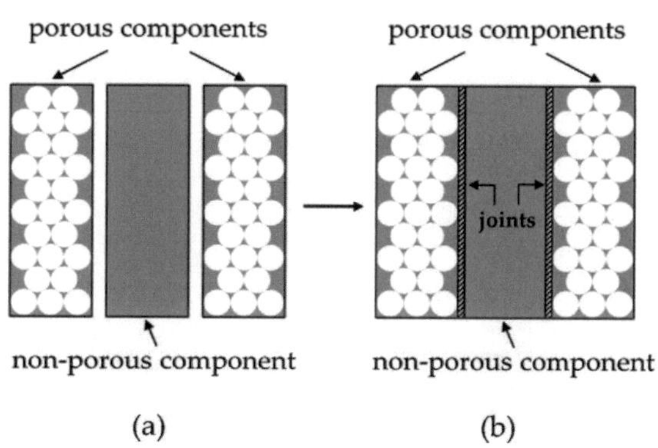

Figure 4.
Schematic drawings showing component composition (a) and the structure of the final material after components joining (b) for finned foams.

3.2.3 Finned foams

Finned foams are normally manufactured by physical bonding or by gluing preexistent monolith layers (herein called components) of porous and nonporous materials. The nonporous materials are considered the new phases which are integrated into a material with a layered structure and a noncontinuous matrix (joints are present in between components) (**Figure 4**).

3.2.4 Monolithic finned foams

Another way to fabricate finned foams is to perform a casting of a liquid precursor into a mold where preexistent self-standing porous leachable preforms are conveniently located. As a result, monolithic finned foam materials with continuous matrix can be obtained (**Figure 5**).

3.2.5 Composite finned foams

Loading of new phases is achieved by building an assembly consisting of packed or self-standing porous leachable preforms alternated with packed beds of the new phases in finely divided form (inclusions). After infiltration and removal of the leachable materials, a final material with a layered distribution of components and continuous matrix is obtained (**Figure 6**).

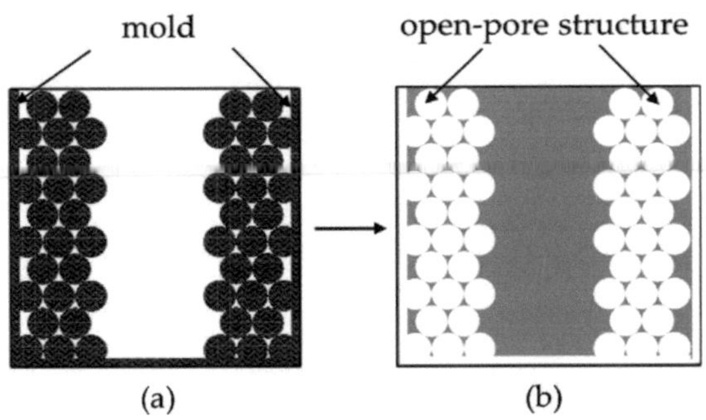

Figure 5.
Schematic drawings showing a mold with preexistent self-standing porous leachable preforms (a) and the structure of the final material (b) for monolithic finned foams.

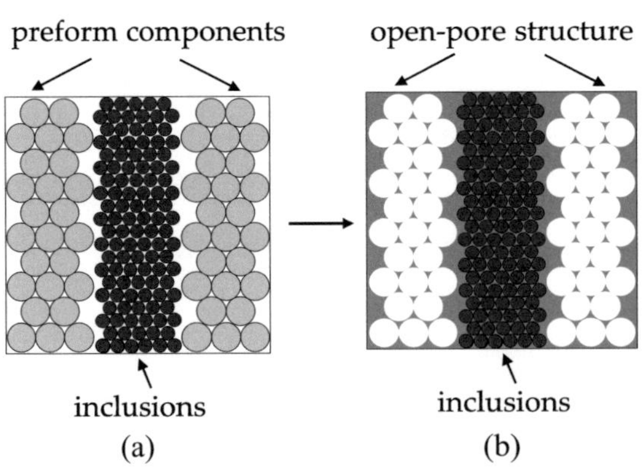

Figure 6.
Schematic drawings showing an assembly consisting of packed or self-standing porous leachable preforms alternated with packed beds of finely divided inclusions (a) and the structure of the final material (b) for composite finned foams.

4. Multiphase open-pore foams: examples and properties

4.1 Composite foams/foams with guest phases with preload of new phases in the preform

4.1.1 Magnesium/diamond composite foams

Open-pore magnesium foams, which have traditionally been discarded for active thermal management due to their low thermal conductivity values, can be appropriate for heat dissipation applications if they incorporate thermal inclusions such as diamond particles coated with a TiC layer of nanometric dimensions. These multiphase open-pore composite foams can be manufactured by the replication method following a strict processing control. First, a correct distribution of the preform components (NaCl and diamond particles) has to be achieved to ensure homogeneity and complete connectivity of the pores after dissolution. For this purpose, the selection of the composition of bimodal particle mixtures has been studied in detail following a predictive method described in [8, 34–37]. The results of these calculations are depicted in **Figure 7a** for the entire spectrum of NaCl particle fraction in the bimodal mixtures. The complete pore connectivity is achieved when the composition of NaCl in the bimodal (NaCl-diamond) mixture falls in the region of interest represented in **Figure 7a**. In this region the large NaCl particles are touching each other, and the smaller diamond particles are filling the voids left by the sodium chloride particles.

Another critical processing step is the proper control of TiC coating on diamond particles, which allows for high thermal conductance at the interface between the diamond particles and the matrix. The scanning electron microscopy (SEM) images in **Figure 8** illustrate some microstructural features of Mg/diamond composite foam. **Figure 8a** shows the diamond particles homogeneously distributed in the struts of the Mg matrix. **Figure 8b** depicts Si and Fe precipitates on diamond surfaces. During the metal solidification, traces of Si and Fe present in the nominal composition of magnesium segregate toward the interface, enhancing together with the TiC coating the magnesium-diamond interfacial thermal conductance.

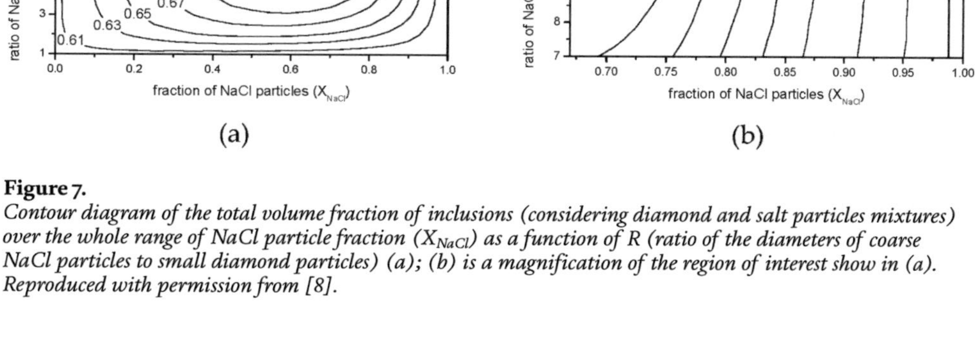

Figure 7.
Contour diagram of the total volume fraction of inclusions (considering diamond and salt particles mixtures) over the whole range of NaCl particle fraction (X_{NaCl}) as a function of R (ratio of the diameters of coarse NaCl particles to small diamond particles) (a); (b) is a magnification of the region of interest show in (a). Reproduced with permission from [8].

Figure 8.
SEM micrographs of TiC-coated diamond particle distribution in the foam struts (a) and fine precipitates of silicon and iron on the TiC-coated diamond particles (b). Sample (a) was prepared by fracture, while sample (b) was prepared by fracture followed by magnesium electro-etching. Reproduced with permission from [8].

The thermal conductivity of these materials was both measured and estimated with analytical methods. The measurement of the thermal conductivity was carried out by the so-called comparative stationary method, which provides accurate and relatively fast measurements. It consists of comparing the thermal conductivity of an unknown material (the sample) with that of a reference, connecting the sections of both and establishing a thermal gradient.

The thermal conductivity was estimated with the differential effective medium (DEM) scheme, which has been extendedly applied with success to model and interpret thermal conduction in different composite materials consisting of randomly distributed monodispersed particles in a metal matrix [8, 34, 35, 37, 38]. The leading equation is expressed as the following integral:

$$\int_{K_m}^{K_C} \frac{dK}{K \sum_i X_i \frac{-\left(K - K_r^{eff}\right)}{\left(K - K_r^{eff}\right)p - K}} = -ln\,(1 - V) \tag{1}$$

where K is thermal conductivity and subscripts C and m refer to composite and matrix, respectively. X_i is the fraction of the i inclusion type in the total amount of inclusions of the composite (in composites containing only one type of inclusions $i = 1$; hence, $X_1 = 1$). V is the total volume fraction of inclusions, and p is the polarization factor of an inclusion (equal to one-third for spheres). K_r^{eff} is the effective thermal conductivity of an inclusion which, for spherical geometries, is related to its intrinsic thermal conductivity, K_r^{in}, the matrix/inclusion interface thermal conductance h, and the radius of the inclusion r, by

$$K_r^{eff} = \frac{K_r^{in}}{1 + \frac{K_r^{in}}{hr}} \tag{2}$$

In general, the integral on the left-hand side of Eq. (1) has no analytical solution and needs to be solved numerically with appropriate mathematical software.

Foam materials can be considered composites where pores are inclusions of zero nominal thermal conductivity ($K_r^{in} = 0$ W/mK). Eq. (1) then becomes

$$K_C = K_{foam} = K_m \left(1 - V_p\right)^{\frac{1}{1-p}} \tag{3}$$

where V_p refers to the volume fraction of pores and p is now the polarization factor of the pores (equal to the polarization factor of the NaCl particles from which the pore structure of the foam was derived, since the replication method maintains the morphological characteristics of the leachable particles in the pores of the final material). For spherical particles again $p = 1/3$ [8]; in more complicated particle geometries, the value of p can be derived from the slope of a plot of $\log(K_{foam})$ vs. $\log(1 - V_p)$ for foam materials in which V_p is varied. K_m is the thermal conductivity of the matrix in the foam material, which can be in turn calculated with Eq. (1) by considering that the matrix is an effective composite material of pure magnesium with diamond particles as thermal inclusions.

The calculated thermal conductivities of Mg/diamond composite foams according to Eqs. (1)–(3) are plotted in **Figure 9a** against the experimental results [8]. In general, large fractions and large average sizes of diamond particles in the matrix generate higher thermal conductivities, and the presence of nano-coated TiC diamond is necessary to overcome the thermal conductivity of magnesium foam, reaching values up to 82 W/mK when the material contains 30% of nano-coated TiC diamond.

Since there is no standardized methodology for testing heat sinks in induced-convection active thermal management, the author of [8] proposed a new experimental setup inspired by that reported in [39] to measure power dissipation density of open-pore materials. Results obtained with this setup showed that composite foams achieved excellent performance in active thermal management with values up to 100% higher than their equivalent magnesium foams and 20% superior than conventional aluminum foams (**Figure 9b**).

4.1.2 Aluminum/graphite flake composite foams

These multiphase foam materials were inspired by the recently developed family of highly anisotropic thermally conductive ternary composites formed by the combination of graphite flakes (Gf), ceramic particles, and a metal matrix [40, 41]. Aluminum/graphite flake (Al/Gf) composite foams combine the appeal of using Gf to improve thermal conductivity with the advantages of metal foams and configure a new family of foam materials with great potential for active thermal management applications. These materials are fabricated using the replication method, replacing the ceramic particles of the ternary composites with

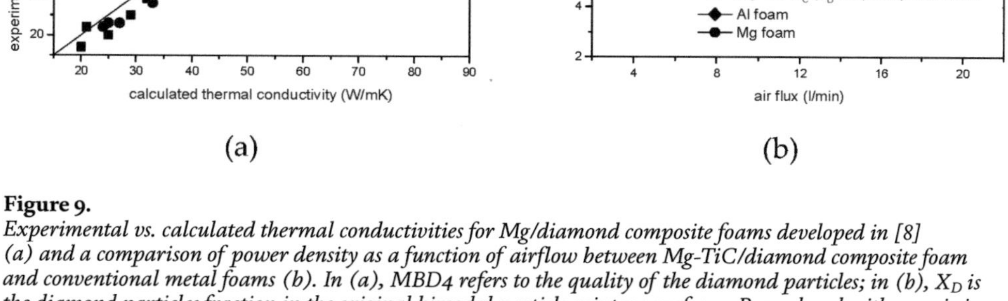

(a) (b)

Figure 9.
Experimental vs. calculated thermal conductivities for Mg/diamond composite foams developed in [8] (a) and a comparison of power density as a function of airflow between Mg-TiC/diamond composite foam and conventional metal foams (b). In (a), MBD4 refers to the quality of the diamond particles; in (b), X_D is the diamond particles fraction in the original bimodal particle mixture preform. Reproduced with permission from [8].

sodium chloride particles, which act as templates and can be removed by dissolution to obtain a material with an interconnected porous structure. The preforms were prepared by packing under external pressure a homogeneous distribution of oriented Gf and NaCl particles.

Two restrictions were found according to the preparation of these preforms. The first one was related with the dissolution of the template. To ensure complete and effective dissolution, the NaCl particles must achieve a coordination number for each particle of at least 3. This restriction defined a so-called percolation limit that is shown in **Figure 10a** as a dotted straight line. The second restriction has to do with the existence of a minimum volume fraction attained when particles are subjected to the sole action of gravity. This second restriction, the so-called compaction limit, is represented in **Figure 10a** by the line corresponding to a nominal zero pressure. As a consequence, preforms with compositions falling in regions below these two limits of **Figure 10a** cannot be manufactured.

For these microstructures where Gf are oriented and distributed homogeneously in a matrix, we can take the following expression for the longitudinal thermal conductivity of composite foams K_C^L [13, 41]:

$$K_C^L = K_{foam} + K_{foam} \frac{V'f}{\frac{\pi t}{4D}(1 - V'f) + \frac{K_{foam}}{K_f^L - K_{foam}}} \quad (4)$$

where K_f^L is the longitudinal thermal conductivity of Gf, $V'f$ is the volume fraction of Gf in the composite material, and t and D are the thickness and diameter of Gf, respectively. K_{foam} can be calculated with Eq. (3).

The calculated vs. the experimental results of longitudinal thermal conductivity for these composite foams are represented in **Figure 11a**. **Figure 11b** depicts the power dissipation densities of two Al/Gf composite foams (one with $V'f = 0.54$ and another one with $V'f = 0.34$) and a conventional aluminum foam (with a porosity volume fraction of 0.78) vs. airflow, obtained with the setup described in [8]. It is clear that proper designs of Al/Gf composite foams can reach power dissipation densities three times higher than those achieved with conventional aluminum foams.

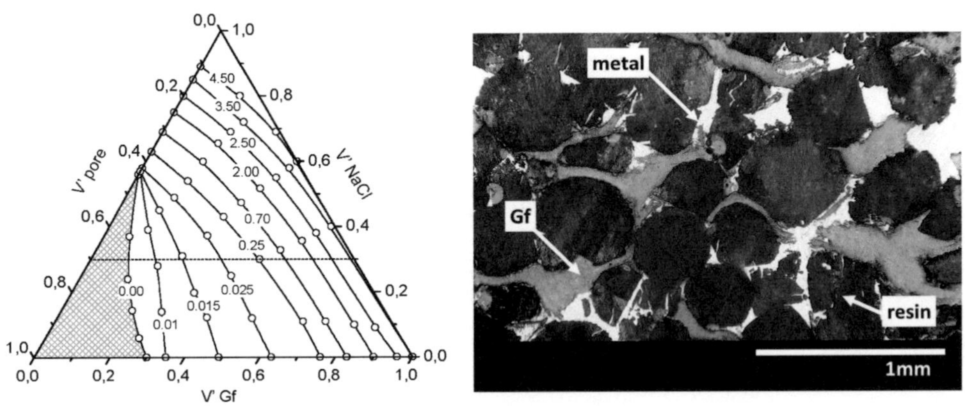

(a) (b)

Figure 10.
(a) Preform composition ternary phase diagram as a function of compaction pressure (in MPa) and (b) photograph of an Al/Gf foam with homogeneous distribution of oriented Gf along the porous material. In (b) the pores were infiltrated with epoxy resin for a better polishing. Reproduced with permission from [13].

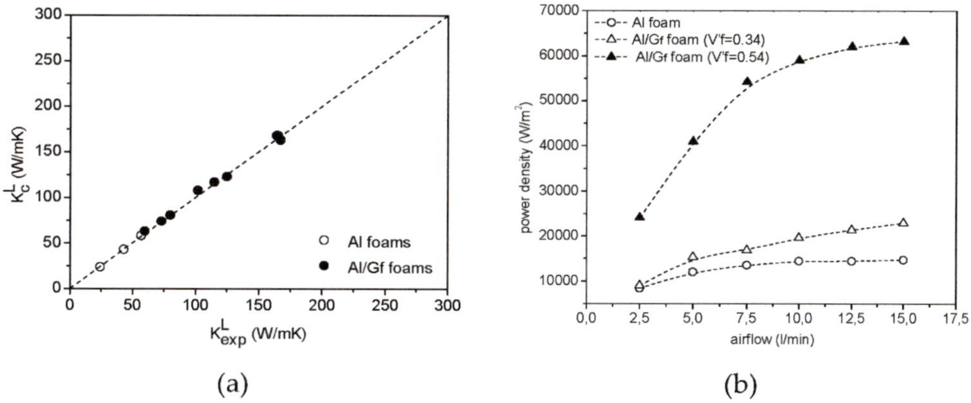

Figure 11.
(a) Calculated vs. experimental thermal conductivities for different Al/Gf composite foams and (b) a comparison of power dissipation density as a function of airflow between some selected Al/Gf composite foams and a conventional aluminum foam. Graphit flakes dimensions: 1000 μm average diameter and 20–45 μm thickness. Partially reproduced from [13].

4.1.3 Foams with guest phases

Open-pore foams containing guest phases in porous cavities is one of the latest developments in the design of foam materials that brings specific functionalities and opens niches for new applications [15, 16]. Depending on the nature of the guest phases or the combination of them, several applications can be considered for these materials such as adsorption of gases, adsorption of liquids or species in solution, catalysis, filters of inorganic or biological substances, and medical implantology. The processing route involves the generation of preforms by packaging particles coated with a sacrificial material (e.g., NaCl), their subsequent infiltration with a suitable precursor, and finally the dissolution of the sacrificial material. This results in open-pore foams in which the cavities contain other phases that provide certain functionalities. A distinctive feature is that there is no bond between the matrix and the guest phases, except for a simple contact generated by gravity, so that the interconnectivity of the pores remains assured. **Figure 12** shows SEM images of a guest particle coated with a sacrificial material (NaCl in this case) (a) and the same guest particle inside the pore cavity of a metal foam (b).

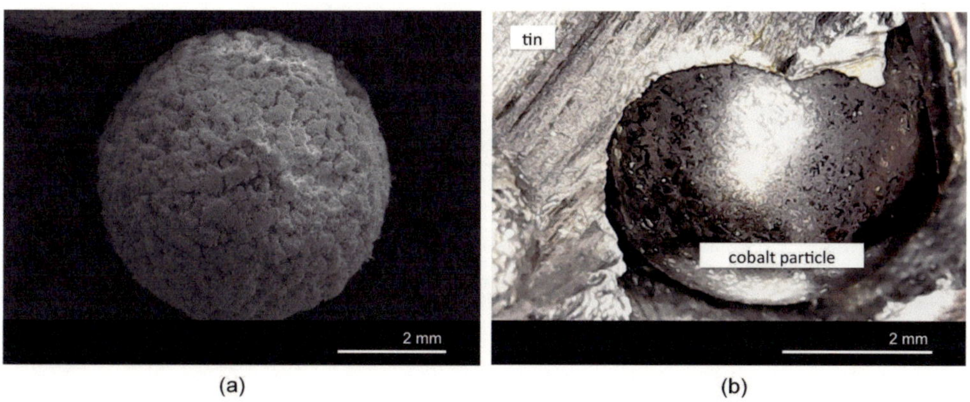

Figure 12.
Cobalt sphere with NaCl coating (a) and cobalt sphere as guest phase inside a cavity of an open-pore tin foam (b). Reproduced from [15, 16].

These materials have not yet been widely characterized, but it is intuited that they have a great potential in the following applications:

- Thermal dissipation by forced convection: the presence of guest phases in the porous cavities alters the distribution of fluid flow lines inside the pores and generates a greater interaction of the fluid with the pore walls, which translates into increased fluid heating and consequently into greater heat dissipation power.

- Catalysis: the material allows catalytically active specimens to be housed in the guest phases and under certain conditions promotes non-laminar regimes in the passage of fluids through it, which notably increases its catalytic activity. In addition, this material can be considered multi-catalytic when different catalytic centers, supported on guest phases physically separated, are combined.

- Medical implantology: the material can act as an implant, allowing the ingrowth of living tissue. The presence of guest phases with adsorbent capacity may be helpful to retain some substances with pharmacological activity that can be released in a controlled manner by desorption and thus avoid possible infections.

4.2 Composite foams with preload of new phases in the liquid precursor

4.2.1 Graphite/TiC nanocomposite foams

TiC-supported metals are interesting systems for catalytic applications. In [14] a new route was presented for the manufacturing of mesophase pitch foam materials containing TiC nanoparticles selectively distributed in two locations (**Figure 13**): in the foam struts (A zone) and at the pore surfaces (B zone). The particles of the struts act as catalysts of the graphitization process to which the mesophase pitch foams are subjected in order to considerably increase their thermal conductivity. The TiC particles on the surface allow transition metals with catalytic capacity to be supported.

As expected, it was found that the higher the TiC content in A zone, the greater the thermal conductivity of these open-pore multiphase foams (thermal

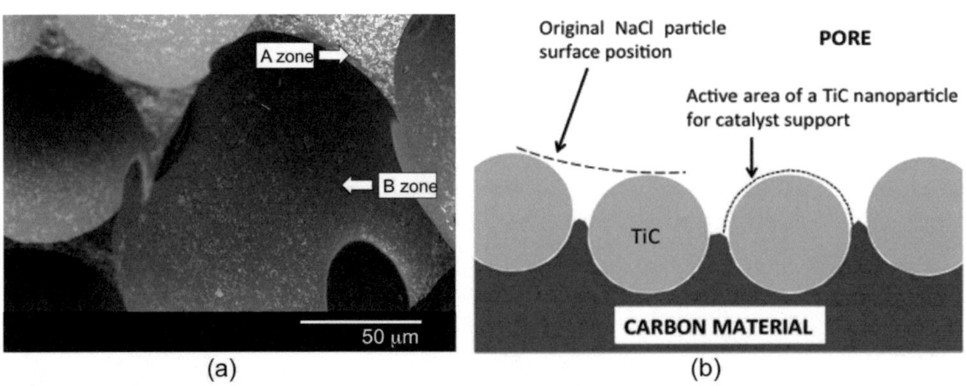

Figure 13.
(a) SEM image showing the location of TiC nanoparticles in the foam struts (A zone) and at the pore surfaces (B zone) and (b) a schematic drawing showing the TiC nanoparticles at the pore surface, which are not completely embedded in the carbon-based material. Reproduced with permission from [14].

conductivities up to 61 W/mK were measured for materials with 15% TiC in A zone and 45% pore surface coverage). The TiC particles at the pore surfaces do not modify the thermal conductivity of the foams, as they are not involved in the graphitization process. However, the higher the nanoparticle content at the pores, the greater the specific surface area of the foam, as the nanoparticles are only partially embedded in the mesophase pitch when infiltration takes place, as can be seen in **Figure 13b**.

4.2.2 Metal/ceramic composite foams

Composite foams are attractive because of their thermal properties, but also because they exhibit interesting mechanical properties when compared to their equivalent raw materials. Many research groups focused their efforts on modifying metal foam microstructures by adding particle reinforcements to enhance their mechanical properties. For that sake, Ni/SiC and Ni/Cu composite foams were proposed in literature [22]. They were manufactured by electrochemical (co)-deposition of the metal and the ceramic particles on polymeric templates. Stainless steel/titanium carbonitrides were also successfully prepared by the replication method [23].

AC3A aluminum alloy/SiC composite foams were manufactured by a similar synthesis route as that described in Section 3.2.2 [20]. The incorporation of the ceramic particles in the foam material strongly improved the compressive strength, energy absorption, and microhardness. The improvement of these properties was due to the modification of the microstructure and the increased strength at the locations where SiC particles were incorporated.

4.3 Finned foams

Finned metal foams were also presented as new designs for thermal applications in [30, 31]. The multiphase open-pore materials developed by Bhattacharya and Mahajan [30] combine alternated parallel aluminum fins with 5 and 20 PPI aluminum foams joined with epoxy glue, as it was previously schematized in **Figure 4**. The results reported in [30] show that these finned foams enhance the heat transfer performance in comparison with conventional aluminum foams, being this increase proportional to the number of fins in the foam. Compared with equivalent aluminum foams, an increase of approximately 150% was reached when fine fins were incorporated. Despite the improved thermal performance of finned foams, the existent joints between components result in poor heat transfer among them. In order to improve the heat transfer between components, new finned foam structures with continuous matrix (no joints) were developed in the last years, and their main characteristics are detailed in next sections.

4.4 Monolithic finned foams

To maximize heat transport between components, monolithic finned copper foams with different geometries of pores were fabricated by a new manufacturing process presented in [33]. 3D printed polymeric or wax patterns were used as sacrificial materials in an investment casting process. This process eliminates the need to restrict design geometries to shapes that can be easily separated from a reusable mold. Their structure, hence, allow these materials to be classified as a combination of monolithic materials with a continuous matrix (Section 3.2.4).

4.5 Composite finned foams

4.5.1 Aluminum/graphite flake composite finned foams

These multiphase materials were also inspired by those presented in [40, 41]. Preforms were prepared by uniaxial pressure packaging of alternating layers of graphite flakes and NaCl particles. Preforms were infiltrated by the gas pressure technique with liquid aluminum and later leached away by water dissolution.

In this type of preforms, there are no restrictions concerning percolation, as the structure could ideally be understood as composed of alternating porous NaCl and Gf monoliths. Even in the extreme case where the percentage of NaCl monoliths is negligible compared to that of Gf monoliths, the NaCl particles in the monolith still have enough coordination to be effectively removed by dissolution. Nevertheless, a compaction limit is detected for preforms prepared without external pressure as a result of the natural tendency of graphite flakes to lie on top of one another (**Figure 14a**). The resulting multiphase open-pore foams present microstructures with alternating layers of oriented graphite flakes and metal foam, as it is shown in **Figure 14b**.

For alternating layers of Al foam and Gf monoliths, the longitudinal thermal conductivity of the composite finned foams Kc^L can be estimated by the well-known Maxwell approach [13, 41]:

$$K_C^L = V'f.K_f^L + (1 - V'f).K_{foam} \tag{5}$$

where the symbols have the same meaning as in Eq. (4) and K_{foam} is again calculated with Eq. (3).

Analytical values obtained from Eq. (5) are correlated with the experimental results in **Figure 15a**. As it can be seen, the model represented by Eq. (5) can reasonably predict the longitudinal thermal conductivities for the Al/Gf composite finned foams, which reach experimental values up to 290 W/mK. The power dissipation density results obtained under working conditions with the setup described in [8] are represented in **Figure 15b**. The experimental results show increments in power dissipation density up to 325% compared with conventional aluminum foams.

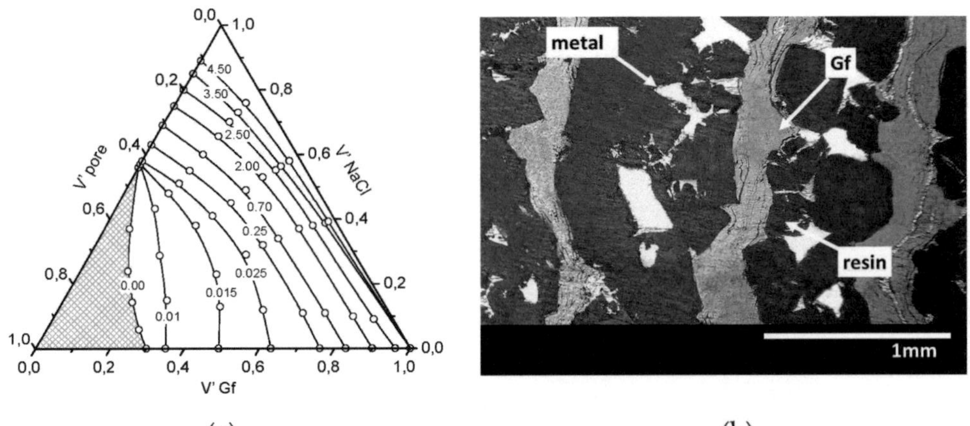

(a) (b)

Figure 14.
(a) Preform composition ternary phase diagram as a function of compaction pressure (in MPa) and (b) photograph of an Al/Gf foam with alternating layers of oriented graphite flakes and metal foam. Reproduced with permission from [13].

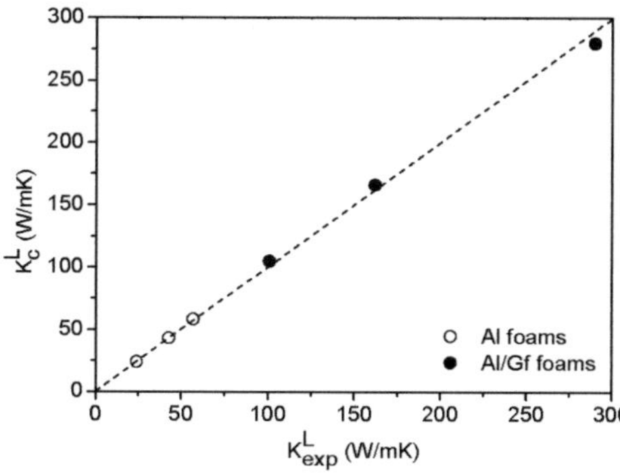

(a)

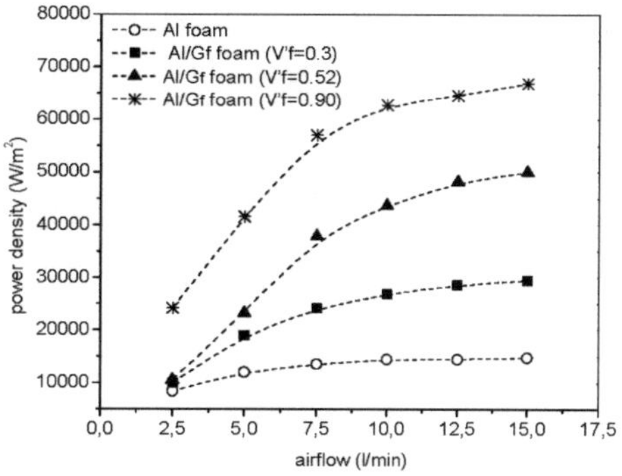

(b)

Figure 15.
(a) Calculated vs. experimental thermal conductivities for different Al foams and Al/Gf composite finned foams and (b) a comparison of power dissipation density as a function of airflow between some selected Al/Gf composite finned foams and a conventional aluminum foam. Graphite flake dimensions: 1000 μm average diameter and 20–45 μm thickness. Partially reproduced from [13].

5. Conclusions

This chapter reviews recent developments in the manufacture and characterization of multiphase foams developed by incorporation of new phases into open-pore foam materials. The new incorporated phases can significantly alter the macro-/microstructure of the starting materials or modify the pore surfaces to achieve new functionalities.

The incorporation of new phases into open-pore foams opens up a new range of properties in foam materials since improvements can be obtained in the mechanical, thermal, catalytic, or adsorptive properties, among others. The design and conception of multiphase open-pore foams seem to be a very suitable way to overcome the growing demands for very specific properties in some modern applications in sectors such as electronics, catalysis, or medical implantology.

Acknowledgements

The authors acknowledge partial financial support from the Spanish Agencia Estatal de Investigación (AEI) and European Union (FEDER funds) through grant MAT2016-77742-C2-2-P.

Conflict of interest

The authors declare no conflict of interest.

Nomenclature

D	graphite flakes diameter (m)
Gf	graphite flakes
h	matrix/inclusion interface thermal conductance (W/m^2 K)
K_C	thermal conductivity of composite (W/mK)

$K_C{}^L$	longitudinal thermal conductivity of composite foam (W/mK)
$K_{exp}{}^L$	experimental thermal conductivity of composite foam (W/mK)
K_f^L	longitudinal thermal conductivity of graphite flakes (W/mK)
K_{foam}	thermal conductivity of foam (W/mK)
K_m	thermal conductivity of matrix (W/mK)
K_r^{eff}	effective thermal conductivity of inclusion (W/mK)
$K_r{}^{in}$	intrinsic thermal conductivity of inclusion (W/mK)
p	polarization factor of inclusion
r	radius of inclusion (m)
t	graphite flake thickness (m)
V	volume fraction of inclusions
$V'f$	volume fraction of graphite flakes
V_p	volume fraction of pores
Xi	fraction of i inclusion type in the total amount of inclusions

Author details

Lucila Paola Maiorano Lauría[1] and José Miguel Molina Jordá[2*]

1 University Materials Institute of Alicante (IUMA), University of Alicante, Alicante, Spain

2 Inorganic Chemistry Department, University Materials Institute of Alicante (IUMA), University of Alicante, Alicante, Spain

*Address all correspondence to: jmmj@ua.es

References

[1] Banhart J. Manufacture, characterisation and application of cellular metals and metal foams. Progress in Materials Science. 2001;**46**:559-632

[2] Hammel EC, Ighodaro OLR, Okoli OI. Processing and properties of advanced porous ceramics: An application based review. Ceramics International. 2014;**40**:15351-15370

[3] Khemani KC. Polymeric Foams: An Overview. Washintong, DC: ASC Symposium Series, American Chemical Society; 2009. pp. 1-7

[4] Singh S, Bhatnagar N. A survey of fabrication and application of metallic foams (1925-2017). Journal of Porous Materials. 2018;**25**:537-554

[5] Lewis G. Properties of open-cell porous metals and alloys for orthopaedic applications. Journal of Materials Science. Materials in Medicine. 2013;**24**:2293-2325

[6] Schelling PK, Shi L, Goodson KE. Managing heat for electronics. Materials Today. 2005;**8**:30-35. DOI: 10.1016/s1369-7021(05)70935-4

[7] Mallik S, Ekere N, Best C, Bhatti R. Investigation of thermal management materials for automotive electronic control units. Applied Thermal Engineering. 2011;**31**:355-362. DOI: 10.1016/j.applthermaleng.2010.09.023

[8] Molina-Jordá JM. Multi-scale design of novel materials for emerging challenges in active thermal management: Open-pore magnesium-diamond composite foams with nano-engineered interfaces. Composites Part A: Applied Science and Manufacturing. 2018;**105**:265-273. DOI: 10.1016/j.compositesa.2017.11.020

[9] Shih WH, Chiu WC, Hsieh WH. Height effect on heat-transfer characteristics of aluminum-foam heat sinks. Journal of Heat Transfer. 2006;**128**:530

[10] Zaragoza G, Goodall R. Development of a device for the measurement of thermal and fluid flow properties of heat exchanger materials. Measurement: Journal of the International Measurement Confederation. 2014;**56**:37-49

[11] Zaragoza G, Goodall R. Metal foams with graded pore size for heat transfer applications. Advanced Engineering Materials. 2013;**15**:123-128

[12] Mancin S, Zilio C, Diani A, Rossetto L. Air forced convection through metal foams: Experimental results and modeling. International Journal of Heat and Mass Transfer. 2013;**62**:112-123. DOI: 10.1016/j.ijheatmasstransfer.2013.02.050

[13] Maiorano LP, Molina JM. Challenging thermal management by incorporation of graphite into aluminium foams. Materials and Design. 2018;**158**:160-171. DOI: 10.1016/j.matdes.2018.08.026

[14] Molina-Jordá JM. Mesophase pitch-derived graphite foams with selective distribution of TiC nanoparticles for catalytic applications. Carbon. 2016;**103**:5-8. DOI: 10.1016/j.carbon.2016.02.051

[15] Molina Jorda JM. Spanish Patent P201730890. 2017

[16] Molina-Jorda JM. PCT Patent PCT/ES2018/070474. 2018

[17] Körner C, Singer RF. Processing of metal foams—Challenges and opportunities. Advanced Engineering Materials. 2000;**2**:159-165. DOI: 10.1002/

(SICI)1527-2648(200004)2:4<159::AID-ADEM159>3.0.CO;2-O

[18] Prieto R, Louis E, Molina JM. Fabrication of mesophase pitch-derived open-pore carbon foams by replication processing. Carbon. 2012;**50**:1904-1912. DOI: 10.1016/j.carbon.2011.12.041

[19] Alizadeh M, Mirzaei-Aliabadi M. Compressive properties and energy absorption behavior of Al-Al_2O_3 composite foam synthesized by space-holder technique. Materials and Design. 2012;**35**:419-424. DOI: 10.1016/j.matdes.2011.09.059

[20] Wichianrat E, Boonyongmaneerat Y, Asavavisithchai S. Microstructural examination and mechanical properties of replicated aluminium composite foams. Transactions of Nonferrous Metals Society of China (English Edition). 2012;**22**:1674-1679

[21] Bouwhuis BA, McCrea JL, Palumbo G, Hibbard GD. Mechanical properties of hybrid nanocrystalline metal foams. Acta Materialia. 2009;**57**:4046-4053

[22] Mikutski V, Smorygo O, Shchurevich D, Marukovich A, Ilyushchenko A, Gokhale A. Open-cell metal-SiC composite foams made by electrolytic code position on polyurethane substrates. Powder Metallurgy and Metal Ceramics. 2014;**52**:545-550

[23] Bakan HI, Korkmaz K. Synthesis and properties of metal matrix composite foams based on austenitic stainless steels—Titanium carbonitrides. Materials and Design. 2015;**83**:154-158. DOI: 10.1016/j.matdes.2015.06.016

[24] Antenucci A, Guarino S, Tagliaferri V, Ucciardello N. Improvement of the mechanical and thermal characteristics of open cell aluminum foams by the electrodeposition of Cu. Materials and Design. 2014;**59**:124-129

[25] Devivier C, Tagliaferri V, Trovalusci F, Ucciardello N. Mechanical characterization of open cell aluminium foams reinforced by nickel electro-deposition. Materials and Design. 2015;**86**:272-278. DOI: 10.1016/j.matdes.2015.07.078

[26] Edouard D, Ivanova S, Lacroix M, Vanhaecke E, Pham C, Pham-Huu C. Pressure drop measurements and hydrodynamic model description of SiC foam composites decorated with SiC nanofiber. Catalysis Today. 2009;**141**:403-408

[27] Wang W, Burgueño R, Hong JW, Lee I. Nano-deposition on 3-D open-cell aluminum foam materials for improved energy absorption capacity. Materials Science and Engineering A. 2013;**572**:75-82. DOI: 10.1016/j.msea.2013.02.032

[28] Sun Y, Burgueño R, Vanderklok AJ, Tekalur SA, Wang W, Lee I. Compressive behavior of aluminum/copper hybrid foams under high strain rate loading. Materials Science and Engineering A. 2014;**592**:111-120

[29] Luo Y, Yu S, Liu J, Zhu X, Luo Y. Compressive property and energy absorption characteristic of open-cell SiCp/AlSi9Mg composite foams. Journal of Alloys and Compounds. 2010;**499**:227-230. DOI: 10.1016/j.jallcom.2010.03.172

[30] Bhattacharya A, Mahajan RL. Finned metal foam heat sinks for electronics cooling in forced convection. Journal of Electronic Packaging. 2002;**124**:155

[31] Feng SS, Kuang JJ, Wen T, Lu TJ, Ichimiya K. An experimental and numerical study of finned metal foam heat sinks under impinging air jet cooling. International Journal of Heat and Mass Transfer. 2014;**77**:1063-1074. DOI: 10.1016/j.ijheatmasstransfer.2014.05.053

[32] Wang J, Kong H, Xu Y, Wu J. Experimental investigation of heat transfer and flow characteristics in finned copper foam heat sinks subjected to jet impingement cooling. Applied Energy. 2019;**241**:433-443. DOI: 10.1016/j.apenergy.2019.03.040

[33] Krishnan S, Hernon D, Hodes M, Mullins J, Lyons AM. Design of complex structured monolithic heat sinks for enhanced air cooling. IEEE Transactions on Components, Packaging and Manufacturing Technology. 2012;**2**:266-277

[34] Molina-Jordá JM. Design of composites for thermal management: Aluminum reinforced with diamond-containing bimodal particle mixtures. Composites Part A: Applied Science and Manufacturing. 2015;**70**:45-51. DOI: 10.1016/j.compositesa.2014.12.006

[35] Molina-Jordá JM. Nano- and micro-/meso-scale engineered magnesium/diamond composites: Novel materials for emerging challenges in thermal management. Acta Materialia. 2015;**96**:101-110. DOI: 10.1016/j.actamat.2015.06.003

[36] Molina JM, Saravanan RA, Arpón R, García-Cordovilla C, Louis E, Narciso J. Pressure infiltration of liquid aluminium into packed SiC particulate with a bimodal size distribution. Acta Materialia. 2002;**50**:247-257

[37] Molina-Jordá JM. SiC as base of composite materials for thermal management. In: Mukherjee M, editor. Silicon Carbide— Materials, Processing and Applications in Electronic Devices. Rijeka: IntechOpen; 2011. pp. 115-140. DOI: 10.5772/57353

[38] Molina JM, Narciso J, Weber L, Mortensen A, Louis E. Thermal conductivity of Al-SiC composites with monomodal and bimodal particle size distribution. Materials Science and Engineering A. 2008;**480**:483-488. DOI: 10.1016/j.msea.2007.07.026

[39] Hsieh WH, Wu JY, Shih WH, Chiu WC. Experimental investigation of heat-transfer characteristics of aluminum-foam heat sinks. International Journal of Heat and Mass Transfer. 2004;**47**:5149-5157. DOI: 10.1016/j.ijheatmasstransfer.2004.04.037

[40] Prieto R, Molina JM, Narciso J, Louis E. Fabrication and properties of graphite flakes/metal composites for thermal management applications. Scripta Materialia. 2008;**59**:11-14. DOI: 10.1016/j.scriptamat.2008.02.026

[41] Prieto R, Molina JM, Narciso J, Louis E. Thermal conductivity of graphite flakes-SiC particles/metal composites. Composites Part A: Applied Science and Manufacturing. 2011;**42**:1970-1977. DOI: 10.1016/j.compositesa.2011.08.022

Chapter 2

Thermal and Acoustic Numerical Simulation of Foams for Constructions

Marco Caniato, Giada Kyaw Oo D'Amore and Jan Kašpar

Abstract

Cellular foams are widely employed as insulation materials, both thermal and acoustic, often in competition with traditional fibrous insulation material, e.g., rock wool. As for the acoustic and thermal properties, several models have been developed to predict acoustic properties of poroelastic materials, but they are usually applied to fibrous layers or polyurethane foams, whereas their application to new materials like complex cellular foams has not been assessed due to the different cell microstructures. There is a very strong interest both in industrial and academic in developing novel insulation materials; accordingly, the possibility of ideally designing the cellular foam microstructure to achieve desired acoustic performances appears a highly attractive target. The paper will first discuss the state-of-the-art acoustic and thermal models and their application to cellular foam materials. Then a novel sustainable alginate-based foam material will be analyzed as a case study, by focusing the aspects related to their microstructure and acoustic properties. For the derivation of an acoustic model, the determination of the parameters of Johnson-Champoux-Allard (JCA) acoustic model (tortuosity, viscous characteristic length, thermal characteristic length, porosity, and flow resistivity) was performed using five different forecasting methods, including traditional analytical model for fibrous materials as well as inverse procedure.

Keywords: sustainable cellular foam, acoustic properties of foams, thermal properties of foams, JCA acoustic model, TMM acoustic model, alginate foams

1. Introduction

Synthetic cellular foam materials have been developed in the late 1940s of the last century, whereas mass production of polymeric, mostly polyurethane, foams started a decade later [1]. There is a large variety of application of these materials, ranging from lightweight structures to insulation, thermal, acoustical, filtering applications, etc. [2, 3]. Consistently, about 10% of the annual production of polymers is dedicated to produce foams, highlighting both technological and market importance of these materials. As shown in **Figure 1**, cellular foams have also attracted a still increasing attention of the researchers over the past 50 year or so, totaling over 6300 papers. Equally increasing interest is manifested by industrial researcher, and in line with the technological importance of these materials, the

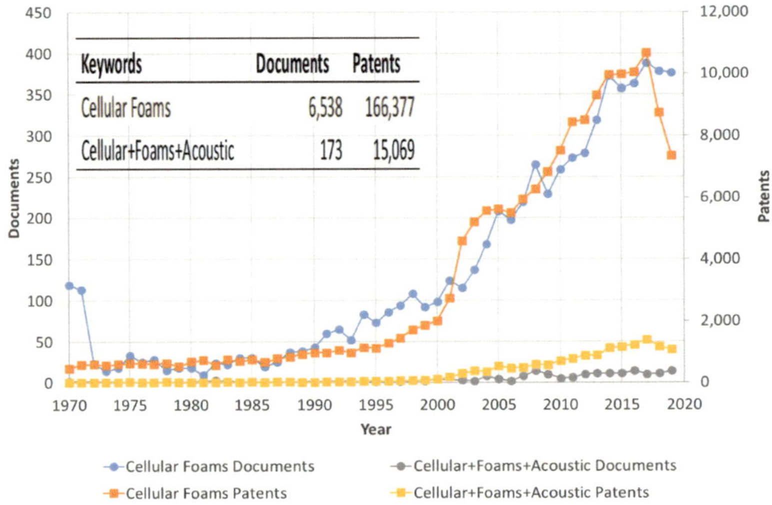

Figure 1.
Results of Scopus search using combination of the indicated keywords. Totals of documents/patens published in 1970–2019 period are also reported (search conducted in December 2019, the lines are used only as eye-guide). Figure adapted from [15].

ratio of published patents/patent applications to the published documents is about 25. It is important to evidence that when the term "acoustic" is added to the previous search, the total drops by about 90%, and most importantly, the above quoted ratio increases up to 87, despite the fact that cellular foams, particularly those with open cells, are widely employed as acoustic insulators [4]. Clearly, despite the technological and market importance of these materials as acoustic insulators, the acoustic aspects and performances of these materials have sparingly been addressed in the scientific literature.

Among the cellular foam materials, polyurethane foams are widely employed, especially in the building sector [5], due to their quite low thermal conductivity (0.022–0.028 W/(m K)), which makes them one of the materials of choice in the applications that require effective and lightweight insulation. In the recent years, circular economy has become a priority in EU countries. Consistently, recycling of waste into industrial products has become an important issue [6]. As far as foam materials are concerned, whereas elastomers or similar waste residues could be effectively recycled in a polyurethane foaming process or even bio-based binders to form efficient acoustics absorbers [7, 8], reutilization of glass and ceramic waste generally employs high-energy-demanding production process [9–11], often leading to foam materials to be employed as insulators. An important class of materials, difficult to be recycled, is represented by composites such as fiber-reinforced thermoset polymers, used in a wide range of industrial application, from automotive to industrial, transportation and naval sectors, etc. [12, 13]. Due to landfill restrictions, thermal (energetic recovery, recovery of fibers/chemicals), mechanical (filler materials), and chemical (solvolysis) routes are underdevelopment, yet their industrial applications still represent a burden, both economic and technical [13]. Recently, we have developed a novel process to recover glass and fiberglass waste via low-temperature foaming process using natural alginate-based foaming agent as a novel route leading to sustainable insulating materials, with interesting acoustic absorption properties [14, 15]. An interesting aspect of this research was linked to the applicability and the limitation of literature acoustic models when used to describe the behavior to this novel cellular type of materials. Specifically, the focus is made on the interconnection between the material properties as obtained from the microstructural characterization and the parameters of Johnson-Champoux-Allard (JCA) acoustic model (tortuosity, viscous characteristic length, thermal characteristic length, porosity, and flow resistivity) [16, 17].

2. Cellular foams: modeling of thermal properties

Due to their intensive use as insulating materials, thermal properties of cellular foams have been investigated quite intensively [18], and a number of models was reported, first of these date back to the 1930s of the last century [19]. Here we summarize only some of the models adopted.

Placido et al. [19] developed a predictive model which considers that heat transfer takes place by both conductions through solid skeleton and included gas and by radiation across the whole layer. The radiation is attenuated by material microstructures, via scattering and absorption phenomena. As for the contribution of the free convection the heat transfer, it is generally considered negligible due to the very small pore size so that the Raleigh number is much less than the critical value [20, 21].

The one-dimensional thermal conduction occurring in a continuum layer is regulated by the Fourier law:

$$q_c = -\lambda \frac{dT}{dx} \tag{1}$$

where q_c is the flux of the thermal conduction (W/m^2), λ is the thermal conductivity (W/(m K)), T is the temperature (K), and x is the thickness (m). The thermal conductivity in foams results as the sum of bulk thermal conductivity and gas one (λ_g). The radiated effect can be modeled by a diffusive equation as follows:

$$q_r = -\lambda_r \frac{dT}{dx} \tag{2}$$

where q_r is the flux of the thermal radiation (W/m^2) and λ_r is the radiative conductivity (W/(m K)) as follows:

$$\lambda_r = 16\, \sigma_{\text{Bolz}} \frac{T_m^3}{3\, K_r} \tag{3}$$

where σ_{Bolz} is the constant of Stefan-Boltzmann, T_m is mean temperature, and K_r is the Rosseland coefficient.

As per the conservation law, the final heat flux and the final conductivity will be

$$q_{\text{tot}} = q_c + q_r \tag{4}$$

$$\lambda_{\text{tot}} = \lambda + \lambda_g + \lambda_r \tag{5}$$

For the radiative part, Cunsolo et al. [22] analyzed several analytical methodologies, concluding that those procedures provide results that are less reliable than numerical ones. This difference is basically caused by porosity modeling, while cell size distribution may not affect final outcomes.

Mendes et al. [23] predict the effective thermal conductivity by means of finite volume methods, taking into account both regular cells and real ones.

Klett et al. [24] demonstrated the paramount effect of rigid skeleton compared to included gas.

Öchsner et al. [18] provided a very comprehensive analysis of thermal property simulation and prediction of porous media, highlighting the difference in approaches both from analytical and numerical point of view. Furthermore, they explain how there are three main theoretical approaches:

i. Field approach, where the Laplace equation is solved taking into consideration the microstructure of the system and the influence of the structural elements in the linear propagation of the flux

ii. Resistor approach, where the bulk and gaseous phases are considered like parallel parts which are assumed to be thermal resistors to flux propagation

iii. Phase averaging, where the effective thermal conductivity is obtained by averaging the constituting phases

Many models were built on cells micro–macro structures like Pande et al. [25] focusing on the heat flow propagation direction [26] or on cell distribution [27].

The Monte Carlo approach is used several time in order to compute temperature profiles where there are randomly shaped borders with varying boundary conditions [28].

Yüksel [29] presented a comprehensive review of measurement methods for the determination of thermal conductivity of materials highlighting how for porous ones only heat flow meter and guarded hot plate are useful to this aim.

3. Cellular foams: modeling of acoustic properties

As indicated above, the acoustic behavior of cellular foams, which is the focus of this work, received less attention compared to thermal applications. The acoustic efficiency of the foams can be easily understood when the principles of sound absorption are considered [30, 31]. The incident sound energy (E_i) interacts whit the material according to Eq. (6) where E_r, E_a, and E_t represent, respectively, energy that is reflected, absorbed, and transmitted:

$$E_i = E_r + E_a + E_t \tag{6}$$

Clearly, as illustrated in **Figure 2**, in in order to minimize transmitted energy, both absorption and reflection must be maximized for our material.

Sound absorption occurs in porous materials essentially via three mechanisms, i.e., (i) interaction of air molecules which vibrate and interact with the pore walls; (ii) air compression and expansion in the pores induced by the entering sound wave, resulting in sound energy transformation into heat; and (iii) vibration and resonance of pore walls [16, 32, 33]. Clearly, all the three mechanisms involve the air located within the pores and its motion that lead to transformation and dissipation of the original sound energy.

From a material point of view, sound absorption coefficient (α) is used to quantify the efficiency of the porous sound absorption materials, which can be measured by impedance tube or reverberation chamber by considering its definition in terms of energy:

$$\alpha = 1 - \frac{E_r + E_t}{E_i} = \frac{E_a}{E_i} \tag{7}$$

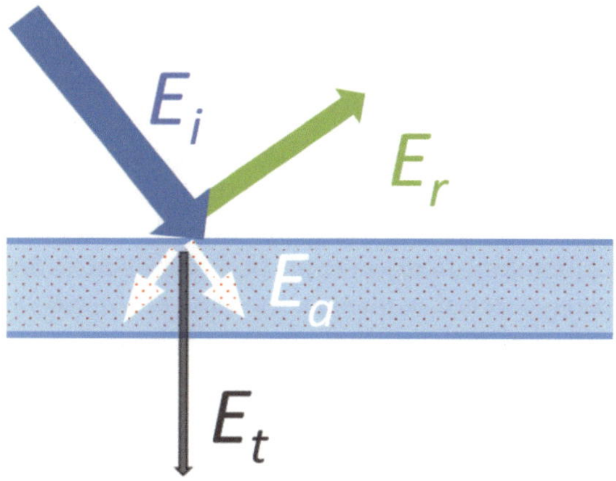

Figure 2.
Scheme of sound energy interaction with a sold: energy conservation [32].

Sound absorption coefficient can be calculated if the surface impedance (Z_s) is known according to Eq. (8):

$$\alpha = 1 - \left|\frac{Z_s - \rho_0 c_0}{Z_s + \rho_0 c_0}\right|^2 \tag{8}$$

where the term $\rho_0 c_0$, respectively, density and speed of sound, represents the impedance of the air.

The aspects of energy dissipation are associated mainly with viscoelastic phenomena that occur within the rigid material and in the interface between it and the fluid in motion. For this reason, the material can be considered biphasic and can be effectively modeled with the Biot theory [34–36]. Therefore, foams can be defined by an equivalent fluid that features an equivalent density and equivalent bulk modulus according to the following equation:

$$\Delta p + \omega^2 \frac{\rho_{\text{eq}}^{\sim}}{K_{\text{eq}}^{\sim}} p = 0 \tag{9}$$

where $\rho_{\text{eq}}^{\sim}$ and $K_{\text{eq}}^{\sim}$ are the equivalent density and the equivalent bulk compressibility modulus and p corresponds to the acoustic pressure (the tilde symbol ($\sim$) specifies that the associated variable is a complex value and related to the frequency). To assess the Biot model with respect to the investigated material, knowledge of macroscopic parameters such as porosity, flow resistance, etc. is necessary. Thus, the knowledge of these parameters allows acoustic behavior to be modeled. These models may use parameters that are based on the definitions from the theory or experimental measurements [37]. Several techniques are, in fact, used to determine the experimental macroscopic parameters, as discussed in Ref. [37]; however, these measurements are not trivial and often time-consuming.

Consequently, a large number of impedance predictive models for obtaining the sound absorption coefficient have been published [16, 30, 38–40], which can be grossly divided into two groups: empirical and theoretical. The empirical methods were initially developed by applying regression methods to large sets of experimental measurements which clearly links them to the specific material considered [39]. Theoretical methods are based on the physics of the sound propagation in the materials (phenomenological methods) and, desirably, include relationships between microstructure and macroscopic properties [41, 42]. This, in fact, has been the case for polyurethane foams, which, however, typically present a regular and well-defined structure, as exemplified in **Figure 3**. For such materials, the tetrakaidecahedra unit cells, Kelvin cell, which represents packing of equal-sized objects together to fill space with minimal surface area, can effectively be applied to describe the foam microstructure [41, 43, 44]. This bottom-up approach therefore seems limited to specific materials with a very precise and defined morphology of the cells.

Thus, a general approach typically employs a semi-phenomenological model.

For the frequency domain, Johnson et al. [45] proposed a model with arbitrary cell shape of the porous material, while Champoux and Allard [46] derived a model that includes the thermal effects inside the porous medium.

At present, the Johnson-Champoux-Allard model appears as the most effective and reliable model to predict frequency behavior throughout the audible range [39]. This model depends on the following parameters:

- Flow resistivity σ (N s m^{-4})
- Porosity ϕ (–)
- Tortuosity α_∞ (–)
- Viscous characteristic length Λ (μm)
- Thermal characteristic length Λ' (μm)

The JCA model has been further modified by researchers with the aim of improving some aspects: Pannetton [47] considered aspects related to the limp frame, and further modification was incorporated in the Kino's models [48]. To our knowledge, successful application of these models is essentially limited to "simple" foams, such as polyurethane ones, as above quoted.

Foams consist mainly of air interrupted by a very thin solid matrix that constitutes the air cells, which leads to broadband sound absorption properties.

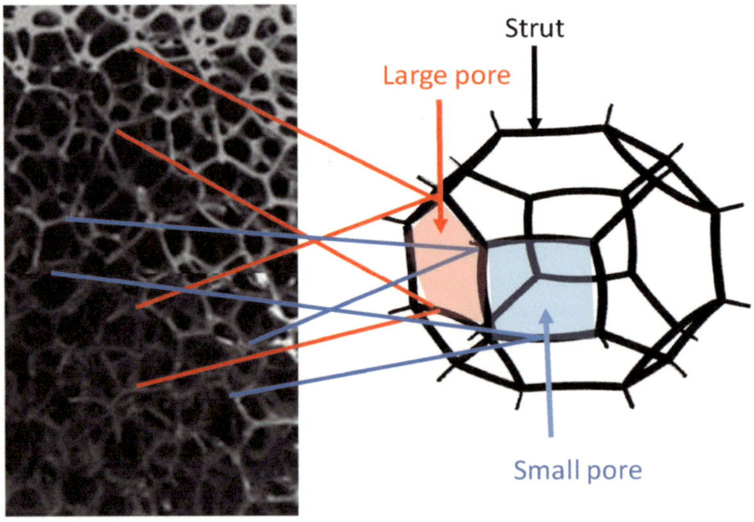

Figure 3.
SEM pictures of two samples of polyurethane (PU) foam and the relationship with the derived model tetrakaidecahedron unit cell (Kelvin cell) of the PU structure (adapted from [44]). Notice the continuity of the cell microstructure across the figure that includes two samples.

Composite-made foams as those here investigated present a complex structure where the alginate matrix is loaded with the glass-containing powder and thus represents a new class of sustainable materials. Given their composite microstructure, it is of strong interest to assess whether traditional numerical acoustic procedures can be effectively used for describing and forecasting their properties.

3.1 Acoustic numerical simulation procedure

In the following paragraphs, the acoustic procedures employed in the present chapter are described. The use of the chosen analytical models is justified by the fact that there are the most used and simple predicting equations found in literature. Thus, an attempt to understand if they could work also with complex foams is important. For detailed description of the experimental details, we refer the reader to our recent publication [15]. The results of this procedures performed on the innovative cellular foams here employed are described in the next section.

3.1.1 Acoustic model parameters

From an analytical point of view, the JCA model parameter, i.e., flow resistivity (σ), porosity (ϕ), tortuosity (α_∞), and characteristic lengths (Λ, Λ'), can be determined using Eqs (10)–(14) [49]. These equations are derived by considering some general assumptions such as periodicity of the microstructures, interconnected pore structures, and small Knudsen number values (ratio of the gas molecular size to a characteristic pore size):

$$\phi = 1 - \frac{\rho_m}{\rho_b} \tag{10}$$

$$\alpha_\infty = 2 - \phi \tag{11}$$

$$\Lambda = r_0 \frac{\phi(2-\phi)}{2(1-\phi)} \tag{12}$$

$$\Lambda' = r_0 \frac{\phi}{1-\phi} \tag{13}$$

where ρ_m is the overall density of the sample and ρ_b is the bulk material composing the open cell.

For air flow resistivity, the most used model is Tarnow's one [50]:

$$\sigma = \frac{4\pi\eta}{b^2(-0.64\ln(d) - 0.737 + d)} \tag{14}$$

where η is the air viscosity, b is the square root of area per fiber, and d is the volume concentration of cylinders.

The acoustic model parameters were calculated form Eqs. (10)–(14) using experimental data (SEM measurements, etc.) and then employed as input values for the transfer matrix method (TMM) [51] calculation of the acoustic absorption coefficient.

3.1.2 Acoustic indirect method

As discussed above, the measurement of the above quoted fluid-phase parameters may be difficult to be obtained and time-consuming. The inverse identification

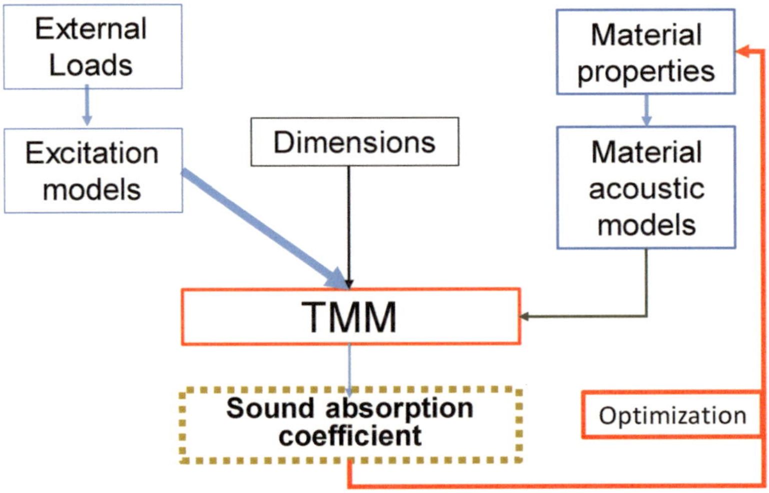

Figure 4.
Optimization TMM functional scheme.

methods fit the acoustical experimental data obtained in a standing wave tube to, i.e., acoustic absorption coefficient as a function of frequency, to calculate the fluid-phase parameters.

Accordingly, the five parameters related to the fluid phase were determined by applying an inversion procedure algorithm described in [52, 53] to experimental laboratory acoustic measurements.

The fitting of the experimental data is based on a nonlinear best-fit approach implemented in the ICT_MAA software (http://www.materiacustica.it/mat_Software_ICT.html).

3.1.3 Acoustic TMM numerical simulation

TMM was used to implement a Johnson-Champoux-Allard model. The general scheme of the TM method is depicted in **Figure 4** where the matrix approach allows introduction of dedicated models according the needed and contemporarily solved.

Eq. (15) reports the general analytical expression for TMM which is normally considered as a two-dimensional problem which considers the impact of a flat acoustic wave on the surface of a structure consistent of two or more layers:

$$V(S_1) = [T]\ V(S_2) \tag{15}$$

The vector $V(S_1)$ contains the variables that define the acoustic indicators (pressure, stresses, velocity, etc.) applied to the surface S_1, whereas the vector V (S_2) contains the same variables for the surface S_2. The matrix T is a function of the physical and mechanical parameters associated with each specific layer.

Accordingly, the transfer matrix [T] models the transmission of sound waves through the layered structure. The dimension of the matrix is a function of the type of the layer, i.e., solid, fluid, poroelastic, or viscoelastic.

Assuming hard-wall boundary condition, i.e., the layered structure being immersed in a semi-infinite fluid on both sides, the complex reflection coefficient can be defined as follows:

$$R = \frac{Z_s \cos\theta - Z_0}{Z_s \cos\theta + Z_0} \tag{16}$$

where $Z_0 = \rho_0 c_0$ represents the characteristic impedance of the fluid calculated my multiplying the density ρ_0 and speed of sound c_0. θ is the incidence angle assumed to be equal to zero, for sound-absorbing coefficient measured at normal incidence. Z_s is the surface impedance of a layer of the package considered calculated as follows:

$$z_s = \frac{\det[D_1]}{\det[D_2]} \tag{17}$$

and

$$\alpha(\theta) = 1 - |R^2| \tag{18}$$

D_1 and D_2 matrices are obtained from a complete matrix D (combination of transfer matrix of each layer, coupling matrices, and proper boundary conditions) and $\alpha(\theta)$ the sound-absorbing coefficient calculated for any θ angle.

4. Cellular foam from recycled waste: synthesis, microstructure and material properties

Generally speaking, cellular foams being porous materials find a large variety of applications, irrespectively of their nature, wherever a lightweight porous material is needed [3], applications as thermal and acoustic insulators being perhaps those most important [54]. Ceramic or glass foam synthesis is traditionally carried out by three routes: (i) replica technique, (ii) use of sacrificial template, and (iii) use of direct foaming agents [2, 55]. There is a common strategy for the first two routes of preparing a precursor of the porous structures at a low temperature. This can be achieved either by impregnation of a "spongelike" material or by using sacrificial particles incorporated in the precursor network. In a subsequent heating step, the sacrificial material is removed, leaving the porous cellular microstructure. In principle this allows to design a specific porous network in the low-temperature synthesis step, which then creates a specific skeleton leading to the porous network during the calcination step. Accordingly, porous structures, ranging from microporous and/or mesoporous to macroporous, could be synthesized [56]. Concerning the third route, the foaming agent is added to the starting mixture. Upon calcination, this agent decomposes generating gas bubbles in the melted material, thus creating the porous structure upon cooling [2, 55]. Typical industrially employed foaming agents are carbonates, particularly in the production of ceramic- and glass-based foams [57].

As stated in the introduction section, there is an increasing attention to the sustainability of material production, and effectively, acoustical sustainable materials, either natural or made from recycled materials, are quite often a valid alternative to traditional synthetic materials [58, 59]. However, the reutilization of glass and ceramic waste generally employs high-energy-demanding production process [9–11], which clearly impacts the sustainability of this route. Furthermore, the use of sacrificial reagents for the synthesis as stated above, clearly contradicts the principles of sustainable chemistry, whereas there is an increasing need for sustainability of both processes and products [60].

We have recently reported synthesis of open-cell foams based on room temperature co-gelling of alginates with glass waste as a viable and sustainable process for production of glass-based cellular. Alginates biopolymers have been used mostly in biomedical applications [61, 62], but recently different applications have been reported, e.g., fire retardants and insulation materials [63], membranes [64], and fuel cell applications [65]. The synthesis of these materials generally implies formation of a gel structure as the "foaming" principle. The gen consists of a continuous solid porous network with pores filled with a fluid, water in our case. Removal of the liquid from the gel to achieve the porous solid causes a significant collapse of pores due to liquid surface tension, particularly high in the case of water, leading to the so-called xerogels where more than 80–90% of pores initially present collapse. Accordingly, to conserve the gel porous structure, water removal must be carried avoiding the liquid–gas interphase. Under supercritical conditions there is no interface between the liquid and gaseous phase; thus, during removal of the fluid, pore collapse is prevented, obtaining high surface products called aerogels [66]. Similarly, sublimation does not imply liquid surface tension, and using freeze-drying technique, the so-called cryogels are obtained [67]. Notably directional freezing was used for the synthesis of cryogels [68, 69], and even alginate gels could be prepared with either isotropic or anisotropic pore structure according to the freeze-drying conditions [70].

Since our interest was focused on eco-efficient glass/fiberglass recycling methodologies [14, 71], the use of alginates, i.e., a natural product, represents a route to improve the greenness of the process. We showed that alginates effectively incorporate recycled glass powders in the gelation step. When these materials are subjected to freeze-drying, open-cell foam structures are formed. Since we were interested in the influence of the coarse foam structure on the acoustic properties, we do not add other bonding agents such as aggregators and plasticizers [72–75], which could confer either flexibility of rigidity to our composite materials. However, we can anticipate that even if the specimens are flexible [76], we do not find significant variation of sound absorption properties in the range of frequencies here investigated. Last but not least, even if freeze drying is considered an energy-demanding unit operation, the comparison of the gelation process with the high-temperature industrial foaming processes showed this process being competitive, less energy-demanding, and cost-effective.

Three samples A, B, and C were prepared containing, respectively, 10 and 20% w/v of glass powders and 20% w/v of fiberglass (see Refs. [14, 71] for details of the syntheses). Recycling fiberglass is difficult and costly being a thermoset composite [13, 77]. In contrast when added in the gel synthesis, they contribute in the creation of the pore structure, leading to a possible, environmentally friendly, recycle route. **Figure 5** shows the microstructure of the samples: macroporous open-cell morphology is observed for all the samples, confirming the efficiency of the proposed methodology for preparing cellular foams. As shown in **Figure 5**(**a1**, **b1**, and **c1**), taken at higher magnifications, both glass and fiberglass are well dispersed and engulfed within the cell walls made of the alginate polymer.

A perusal of **Figure 5**(**a–c**) reveals a net change of the shape and alignment of the pores upon varying the nature of the glass-containing materials and its amount (compare also **Table 2**). Samples A and B feature mostly cells of a quadratic/rectangular form. When the amount of glass powder is increased from 10 to 20% w/v, the orientation of the cells is favored and their dimensions increase. Using fiberglass (sample C), unoriented cells are formed with larger pores compared to sample B. **Table 1** summarizes both the microstructure and mechanical properties of the samples, including the reference rock wool.

Particle size distribution (PSD) curves were measured on the starting glass and fiberglass powders reported in Ref. [15] which provide some insight into this change of microstructure. The glass powder consists of smaller particles compared to fiberglass: the PSD peaks at ca. 8 μm which increases to ca. 128 μm for the fiberglass.

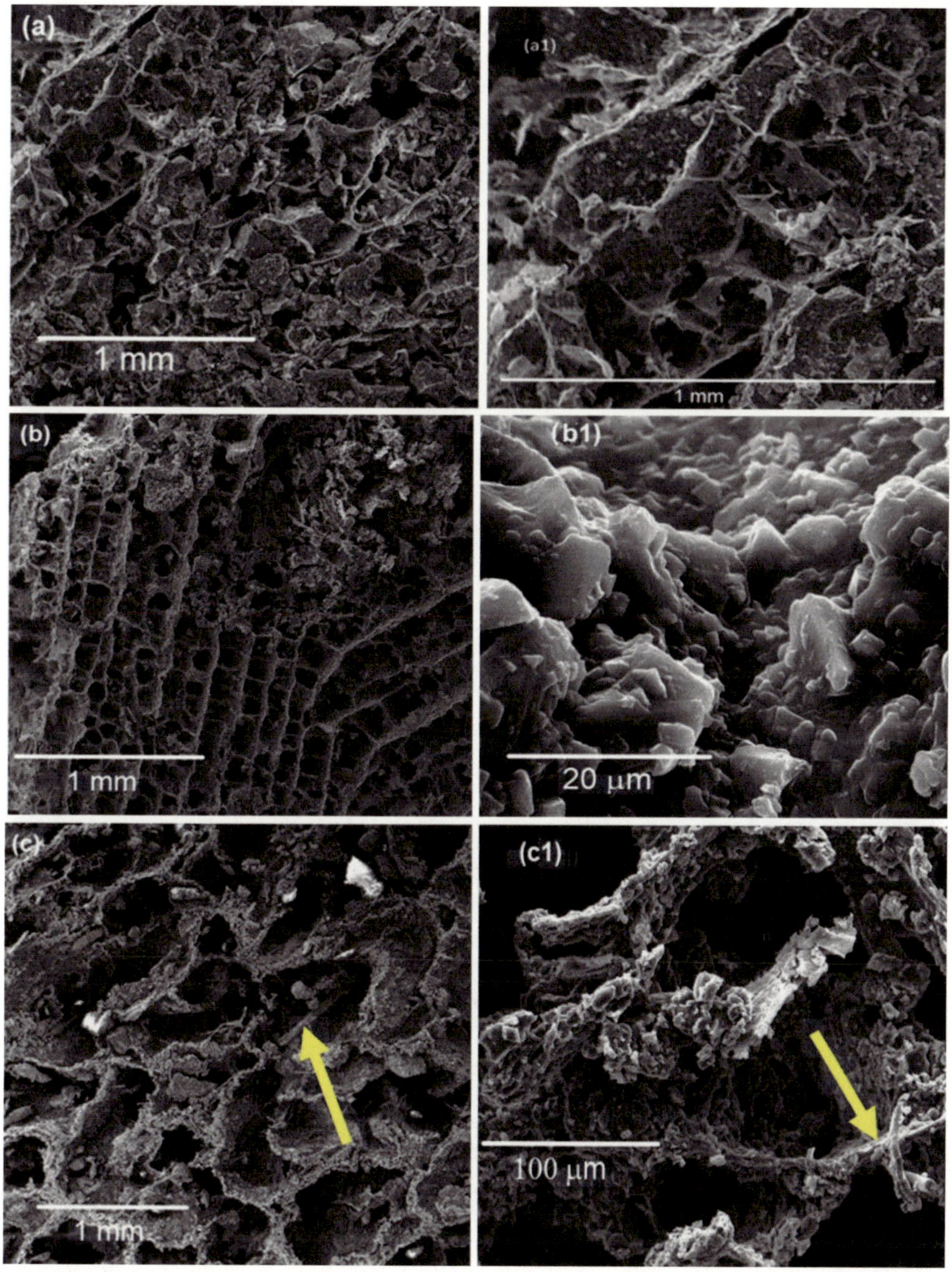

Figure 5.
SEM micrographs: (a) sample A 50×, (a1) details of sample A 50×, evidencing some of the glass powder inclusions; (b) sample B 50×, (b1) details of sample B 2500× evidencing some of the glass powder inclusions; (c) sample C 50×, (c1) details of sample C 500× evidencing some of the fiberglass inclusions. Figure adapted from [15].

Consistently, submillimeter fiber particles, indicated by arrows, are clearly detected in the SEM micrographs reported in **Figure 5(c)**. This change of powder morphology is even more important by considering the particle number (PN) distribution: about 90% of the glass particles are smaller than 4 μm, whereas for an equal percentage, the dimension increases to ca. 60 μm in the case of the fiberglass powder.

Sample	A	B	C	Rock wool
Pore medium area (mm^2)	0.011	0.019	0.074	
Standard deviation (mm^2)	0.011	0.009	0.033	
Radius mean value (μm)	29	38	75	
Porosity	0.85	0.91	0.93	
Density (kg/m^3)	186	201	250	150
E_C (MPa)	5.2	4.2	3.4	1.0
Standard deviation (MPa)	0.6	0.3	0.1	0.1

Table adapted from [15].

Table 1.
Microstructure and properties of the alginate foams: average area and radius of the foam pores, density, and compression modulus.

The process conditions strongly affect freeze-drying synthesis since directional freezing of the ice particles can be easily achieved leading to novel morphologies such as monoliths [69, 78]. This technique can be widely applied, and also alginate-based gels were produced in an anisotropic form [70]. Ordinary freezing conditions were employed for the synthesis, which suggest that this effect should not be operative in our case. It is well-known that during the crystallization of ice, both solute and suspended particles/gels are segregated from the ice crystals. This may generate an ice-templating effect where the morphology of the material is dictated by the crystallized solvent [79]. A large number of small particles favors heterogeneous nucleation providing a large number of nucleation centers [80, 81]. The large amount of small particles in the glass-containing samples A and B increase the ice front velocity promoting formation of a columnar morphology [82], accounting for the morphology detected by SEM. Sample C contains much less small particles, and the rate of nucleation decreases compared to that of particle growth (ice crystallization). This generates an isotropic pattern of the open cells in sample C. The large pore dimension is in line with the higher particle size of the fiberglass compared to glass materials [79, 80].

Thus, the crystallization conditions and the particle distribution in the starting waste material appear to represent factors capable of directing the microstructure leading to distinct cell morphology and dimension. This is an important aspect as the aim of the study is to find correlation between the microstructure and acoustic properties of these materials.

The data reported in **Table 1** show clear trends for the density and the compression modulus which can be correlated with the dimension of the open cells. For a fixed volume, the higher the pore area, the lower the number of cells, which means that the density increases in the sequence samples A, B and C and the opposite occurs for the compression modulus. Data for a rock wool sample are also included in **Table 1**, as a standard sample for the acoustic studies.

5. Cellular foam from recycled waste: acoustic studies

In this section the results of the acoustic performance and application of the different procedures to model the acoustic performance of these novel materials are discussed, first using the analytical procedure to calculate the model parameters and then using the TMM approach.

5.1 Analytical model and acoustic performance

Experimentally measured acoustic absorption coefficient for the samples A, B, and C are reported in **Figure 6**. Samples A and B show comparable shape of the curves where sample B features better global sound-absorbing properties compared to A: the highest absorption coefficient observed for sample B is 0.998 at 2190 Hz. Sample C features a maximum of absorption at about 2100 Hz followed by a slow decline, at variance with samples A and B where a rapid decline is observed. Clearly, different morphologies of sample C compared to A and B lead to different acoustic properties. For comparison, a reference rock wool sample features a nearly linear increase of the sound absorption coefficient with a maximum value of ca. 0.85 at 2900 Hz.

As highlighted above, the application of the analytical model by calculating the JCA parameters using Eqs. (10)–(14) was one of the important aspects of this study. The question is, are the widely employed state-of-the-art parameter formulations applicable to a novel type of material?

To answer this question, we report, for the sake of conciseness, only the result obtained for sample A, but equivalent results have been obtained for samples B and C [15].

In the first instance, in order to model the sound absorption coefficient, the five parameters were calculated according to Eqs. (10)–(14) using the measured densities and the dimensions of the cells evaluated from the SEM micrographs (**Table 2**). As perusal of the data reported in **Table 2** reveals a close similarity of the calculated parameters notwithstanding the dissimilarity in their nature and morphology. This demonstrates that the analytical model is not suitable for this kind of cellular foam microstructure.

The frequency trends of the sound absorption coefficient, which were calculated using these parameters as input for the TMM procedure, are shown in **Figure 6**.

The results (**Figure 6**) show that the analytical model procedure as implemented using Eqs. (10)–(14) cannot be reliably applied to the complex foam structures, at variance with the rock wool sample which is properly modeled. The observation is in line with the above reported comments on the limits of the applicability of this methodology to fibrous materials [48, 50].

The above presented microstructural data show that the morphology and dimensions of the foam cells depend on the addition of the glass-containing powders. Since the powder is incorporated into the walls of the cells, increasing its amount will result in an extension of the free path for the wave propagating within the material itself. As a consequence, a modification of the tortuosity parameter is expected. For this reason, we consider the tortuosity factor as calculated by Eq. (11), developed for fibrous like materials, to inadequately describe this type of novel material. Notice that a modification of the tortuosity parameter changes the sound absorption leaving the thickness of the material unchanged.

Material	Flow resistivity (Eq. (14)) (σ) ((N s) m^{-4})	Porosity (Eq. (10)) (ϕ) (–)	Tortuosity (Eq. (11)) (α_∞) (–)	Tortuosity (Eq. (19)) ($\alpha_{mod,\infty}$) (–)	Viscous characteristic length (Eq. (12)) (Λ) (μm)	Thermal characteristic length (Eq. (13)) (Λ') (μm)
A	24,428	0.92	1.07	2.89	31	57
Rock wool	27,289	0.93	1.06		31	57

Table 2.
Analytical model results: flow resistivity, porosity (ϕ), tortuosity (α_∞), tortuosity $\alpha_{mod,\infty}$, and characteristic lengths (Λ, Λ').

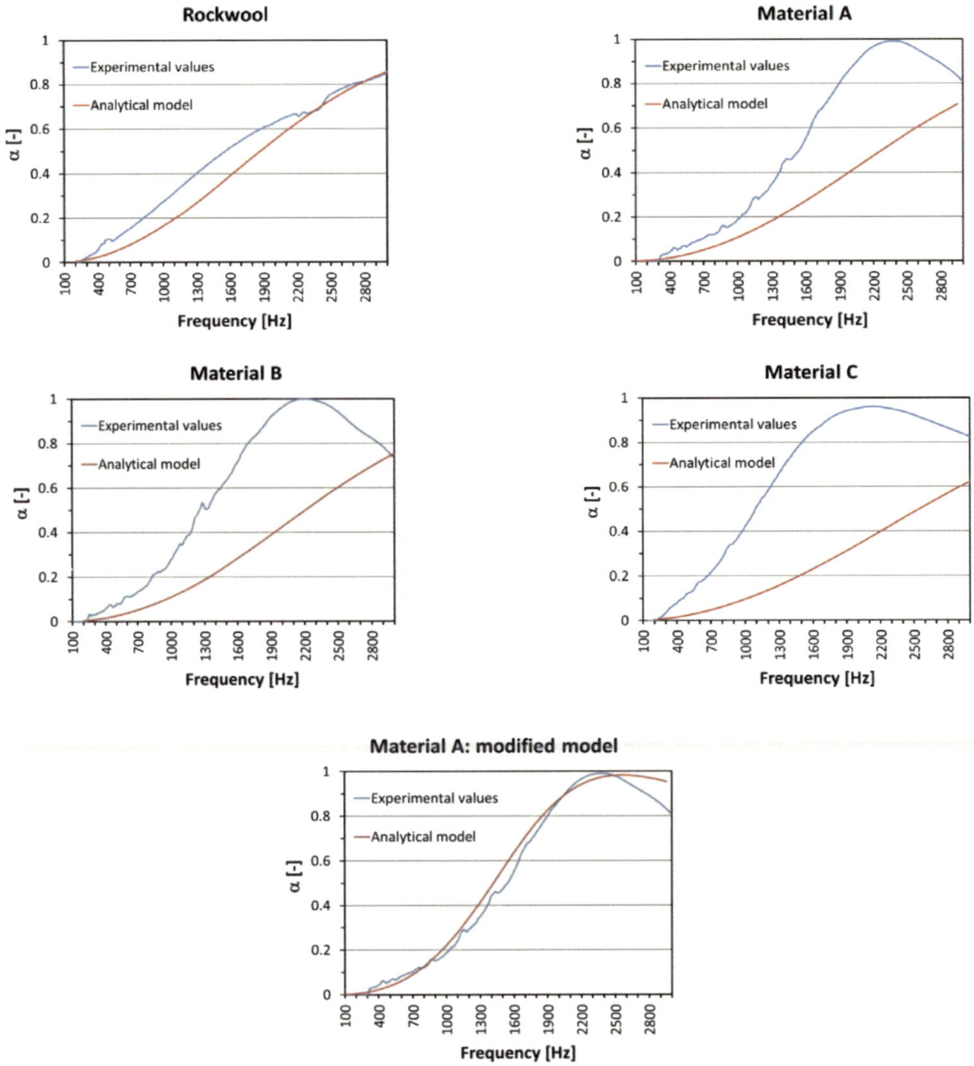

Figure 6.
Sound absorption coefficient as a function of frequency: analytical model (calculated with TMM) vs experimental values obtained for rock wool, samples A, B, and C and sample A using modified tortuosity Eq. (19). Figure adapted from [15].

Eq. (19), which is obtained by modifying the formulation of Archie for the tortuosity [83], is therefore proposed as a partial modification of Eq. (11). Eq. (19) is able to provide a reliable fit, up to 2500 Hz, as shown in **Figure 6**, because this model depends only on the open porosity:

$$\alpha_{\text{mod},\infty} = \frac{1}{\phi^{12.72}} \tag{19}$$

The exponent of the open porosity in Eq. (19) is calculated by a curve-fitting procedure of all measured results. The value of the tortuosity calculated using Eq. (19) is included in **Table 2** for material A. As shown in **Figure 6** (modified model), using the modified tortuosity parameter ($\alpha_{\text{mod},\infty}$) as input, the TMM simulation nicely fits the experimental data.

Accordingly, an important finding of this part of this study is the demonstration of the necessity of adapting the analytical calculation of the parameters for the JCA model to the specific material analyzed. In this case the predictive value of the tortuosity clearly appears strictly related to the nature of the sample. Accordingly, the "traditional" analytical model will not be considered further.

5.2 Acoustic indirect method

As discussed in the preceding section, the modeling of the acoustic properties of porous materials requires to determine physical parameters of the porous solid, namely, airflow resistivity, open porosity, tortuosity, and viscous and thermal characteristic lengths [84]. In the recent years, an inversion method can be applied which consists in a best-fit procedure of the experimental acoustic data to provide all these parameters as the output has become a popular methodology [52]. Such an approach could successfully be applied to a number of different types of porous materials [38]. This is exemplified in **Figure 7** which reports the comparison between the measured and calculated trends for a free inversion of the rock wool sample.

The picture reported in **Figure 7** clearly suggests the effectiveness of this procedure since the modeled data visibly better fit the experimental data compared to the analytical model reported in **Figure 6**. As discussed in Section 4, the final goal of the modeling procedures is to acquire a predictive capacity and, most importantly, the capability to properly correlate the microstructure of the investigated material with its sound-absorbing capacity [42]. This clearly would open new horizons for the material development by trying to develop correlations between the synthesis conditions and material properties [44]. In this respect, it important to recall that the inversion procedure involves a best fit of an experimental curve using a number of parameters, 5 for the JCA model, which can increase up to 8, according to the model considered [30, 85].

The inversion procedure algorithm was therefore applied to the experimental acoustic measurements using three different approaches: in the first one, no restriction has been applied to the inverse procedure. In the second one, restrictions were applied to the values obtained from the modified analytical model. The limitations were applied in terms of upper and lower limits of the flow resistivity (σ) within which the inverse procedure can fit. In the third one, the thermal characteristic length (Λ') value was imposed based on the experimental data (pore radius in **Table 2**) in the inverse procedure. The choice of these restrictions is motivated by

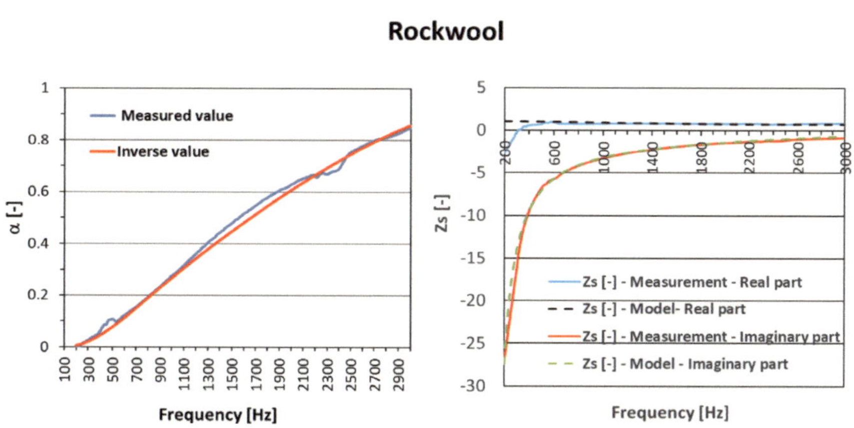

Figure 7.
Comparison of modeled and measured values for rock conditions and wool using parameters obtained from the free inversion procedure. Figure adapted from [15].

the fact that these parameters are those usually experimentally measured in, respectively, acoustic and material science studies.

Figure 8 compares the experimental sound-absorbing coefficient and the complex impedance for the three materials with the calculated, ones using TMM, vs frequency. A quite good agreement between the fitted and experimental curves is found, unrestricted fitting giving the best result for sample A.

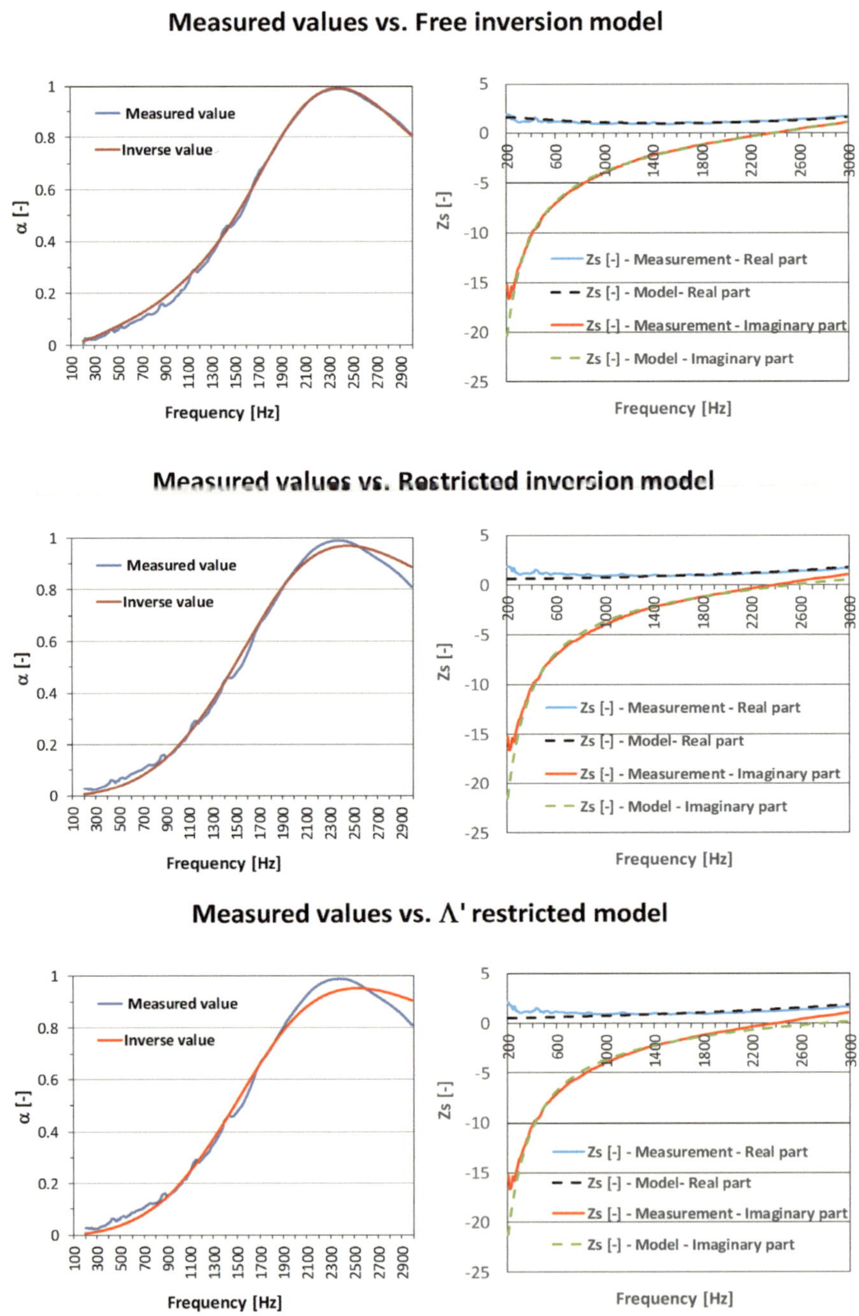

Figure 8.
Comparison of modeled and measured values for sample A using parameters obtained from the inversion procedure, using free inversion, restricted analytical model values and imposing the measured Λ' value: absorbing coefficient vs frequency and complex impedance. Figure adapted from [15].

Material	Flow resistivity (σ) ((N s) m^{-4})	Porosity (ϕ) (–)	Tortuosity (α_∞) (–)	Viscous characteristic length (Λ) (μm)	Thermal characteristic length (Λ') (μm)	Standard deviation[a] (σ)
I. Fitting with no restriction						
A	17,744	0.87	6.78	91	194	0.0142
II. Fitting using restricted method (based on analytical model)						
A	59,676	0.81	4.34	53	53	0.0251
III. Fitting using experimental pore dimension						
A	59,181	0.82	2.88	29	29	0.0323

Table adapted from [15].
[a]Standard deviation of the calculated α from the experimental values assuming that the latter represent the average value of three experiments in the range 200–3000 Hz.

Table 3.
Parameters obtained from inverse procedure using different fitting approaches.

To properly assess the goodness of fit, an attempt was performed using standard deviation calculated as reported in Eq. (20):

$$\sigma = \sqrt{\frac{1}{N}\sum_{i=1}^{N}(x_i - \mu)^2} \tag{20}$$

The calculated values and standard deviation of the calculated α from the experimental values assuming that the latter represent the average value, being an average of three measurements in the range 200–3000 Hz, are reported in **Table 3**.

The values of the calculated deviation shows (i) minor effects of the restriction on the parameters on the goodness of fit upon variation of the fitting procedure, (ii) unrestricted fit is slightly better than those restricted, and (iii) samples A and B are much better fitted than sample C (compare full data in Ref. [15]). As for the latter aspect, this could be related to the irregular pore morphology sample C. Accordingly, the fitting procedure, based on an idealized structure, fits better regular structures compared to those irregular.

6. Comparison of the results

Let us now discuss the four sets of parameter derived by the above described procedures (**Tables 2** and **3**), which were used to calculate the TMM-based forecast of the sound adsorption capability and compared with the experimental data (**Figure 8**).

At first, we observe that by either restricting the inverse fitting procedure or using the parameters calculated in the modified analytical model or imposing the measured Λ' value, the calculated TMM profile fits slightly worse the experimental data compared to those obtained by the unrestricted inverse method.

However, it is important to consider that the derived ϕ, α, Λ and Λ' parameters should be properly related to the real microstructure of foam materials, which should help to discriminate the proper fitting model among of the four considered.

For this purpose, **Figure 9** compares the obtained acoustic parameters for the four sets expressed in terms of relative percentages. For the scope of this paper, we limit the analysis to sample A, being sufficient to provide relevant considerations and insights. For a full comparison of the three samples, we refer the reader to our original paper [15].

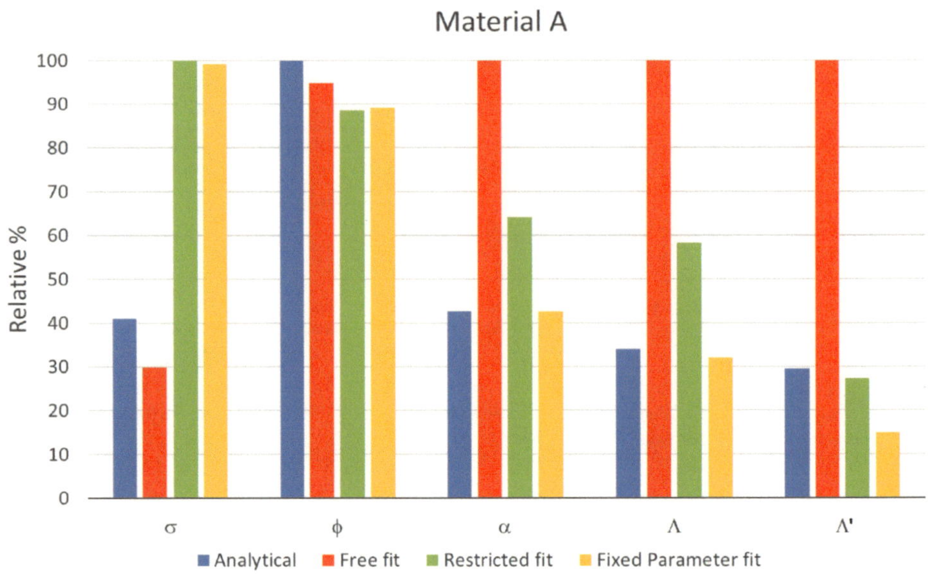

Figure 9.
Comparison of modeled JCA values for sample A with modified analytical, the inversion techniques: values are expressed as % of the maximum value observed for each material/parameter. Figure adapted from [15].

A perusal of **Figure 9** immediately reveals that the "free fit" inverse method computes significantly different values for the α_∞, Λ and Λ' parameters compared to the other fits. Since the calculated values are unrelated to the physical nature of our materials, this is a clear indication that the fit end with a local minimum which, however, has no physical meaning [38, 52].

To be noticed is that the calculated tortuosity (α_∞) shows good agreement for both the modified analytical method and the inversion method with the fixed Λ' parameter. Since Λ' parameter is evaluated from experimental data, this observation confirms the reliability of Eq. (19).

It is worth to remind that conventional materials such as lightweight, fibrous materials (e.g., fiberglass and rock wool) and reticulated foams (e.g., polyurethane and melamine open-cell foams) typically feature porosity and tortuosity very close to unity. In contrast, our and other materials feature tortuosity factors well above unity [38].

The modified analytical methods and fixed Λ' value inverse procedure show a good agreement for ϕ e α parameters, whereas no method accurately estimates the value of the Λ' parameter. Our data indicate that this parameter should be measured experimentally using SEM or an equivalent technique to get a reliable result. This observation highlights direct link between the parameters used in the material science and those used in acoustics.

Both the analytical model and the free inverse fitting (**Table 3**) lead to a low value of the σ parameter compared to the other methods. This however may be explained by the fact that a sensitivity analysis [52] revealed that variation of this parameter scarcely affected the goodness of fit.

The pore geometry is associated with viscous and thermal characteristic lengths [45], the average size of the foam cells being correlated to the thermal characteristic length (Λ'). As for the characteristic viscose length Λ, this parameter, albeit linked to pore geometry, can hardly be derived from the microstructural characterization, whereas its influence is important since narrowing the interconnections between the foam cells, blocks the fluid movement and transition, resulting in improved sound absorption characteristics. As for the similarity of the thermal and viscous characteristic length values, the complex foam cells of our materials feature a parallelepiped interconnected geometry which appears consistent with the similarity of the two parameters.

7. Conclusions

A novel class of sustainable innovative acoustic insulation materials has been described in the present paper. The use of a natural alginate-based gelling agent allows efficient incorporation of waste glass and fiberglass powders. The analysis of the microstructure indicates a strong sensitivity of the pore morphology, on particle dimensions of the doping powder and its amount. The formation of oriented regular cell patterns was attributed to the presence of a large amount of small particles that favors heterogeneous nucleation of ice formation leading to mono-dimensional freezing process. Consistently, using coarse particles produces at comparable doping powder loading an unoriented cellular sample morphology.

Five different forecasting methods including traditional analytical, a modified analytical with a new proposed equation, and inverse procedures were employed to determine the JCA parameters related to the sound-absorbing properties of foam materials. TMM to assess the reliability of the different procedures in comparison to the experimental performance.

The analytical modeling of the JCA parameters, namely, tortuosity, viscous characteristic length, thermal characteristic length, porosity, and flow resistivity showed some limitations of the applicability of the traditional equation, because they are strongly related to fibrous materials rather than foams and a new equation for the determination of the tortuosity was proposed and validated against experimental data using TMM calculation and inverse parameter determination.

The use of the inverse determination of the physical parameters allowed to provide an insight between the materials' properties and acoustic performance: consistent with SEM microstructural analysis indicated comparable foam properties for materials A and B, material C being somewhat different, a situation well consistent with the acoustic performance. As in fact, the sound-absorbing performance depends on cell shape and dimension identified by the thermal lengths. Thus, using the same foaming agent with different doping powders leads to different sound absorption trends: volcano-shaped for materials A and B with glass powder and flat for material C with fiberglass inclusions, as the decline of the sound absorption being less important. The effects of cell orientation impact the acoustic properties as the unoriented cell morphology leads to enhanced sound absorption capacity compared to the samples with more regular and oriented morphology.

An important warning arises from the present data which is the fact that unrestricted fitting may lead to a reliable acoustic profile, corresponding to a local minimum that, however, may not have a physical relationship with the materials properties, e.g., pore morphology. As a matter of fact, the performed sensitivity analysis indicated tortuosity as a factor that heavily affects the fit, which may easily lead physically unreliable values for the other parameters.

Finally, it has been clearly shown that the "traditional" analytical model for determination of JCA parameters cannot be a priori applied to these novel materials due to their complex structure: modification of the calculation of the tortuosity was necessary, and a new equation for the determination of the tortuosity is proposed that has been assessed; the results of the inverse procedure, using the thermal characteristic length derived from the SEM micrographs as imposed parameter, well agree with the modified analytical model. The use of measured values of thermal characteristic length in the inverse procedure is recommended in order to obtain physically reliable results related to the real microstructure. Thus, a direct link between the materials science property and acoustics has been established.

Acknowledgements

This work was financed by "Klimahouse and energy production" in the framework of the programmatic-financial agreement with the Autonomous Province of Bozen-Bolzano of Research Capacity Building, which is gratefully acknowledged.

The authors want to thank Gianluca Turco (Department of Medicine, Surgery and Health Sciences, University of Trieste) for SEM pictures and Andrea Travan (Department of Life Science, University of Trieste) for help in foam production and characterization. Paolo Bonfiglio of Materiacustica srl is gratefully acknowledged for his precious advices.

Author contributions

M.C. developed the research. M.C. elaborated acoustic data, numerical simulation, and acoustic inversions; G.K.d'A. synthetized and characterized the foam samples and performed acoustic measurements with M.C. and J.K. J.K. overviewed the research. M.C., G.K.d'A., and J.K. wrote the paper.

Author details

Marco Caniato[1]*, Giada Kyaw Oo D'Amore[2] and Jan Kašpar[3]

1 Faculty of Science and Technology, Free University of Bozen—Bolzano, Bolzano, Italy

2 Engineering and Architecture Department, University of Trieste, Trieste, Italy

3 Department of Chemical and Pharmaceutical Sciences, University of Trieste, Italy

*Address all correspondence to: marco.caniato@unibz.it

References

[1] Bienvenu Y. Application and future of solid foams. Comptes Rendus Physique. 2014;**15**:719-730. DOI: 10.1016/j.crhy.2014.09.006

[2] Colombo P, Bernardo E. Cellular structures. In: Riedel R, Chen I-W, editors. Ceramic Science & Technology, Structure. Vol. 1. Weinheim, Germany: Wiley-VCH Verlag GmbH & Co. KGaA; 2008. pp. 407-441. DOI: 10.1002/9783527631926.ch10

[3] Wu D, Xu F, Sun B, Fu R, He H, Matyjaszewski K. Design and preparation of porous polymers. Chemical Reviews. 2012;**112**:3959-4015. DOI: 10.1021/cr200440z

[4] Lind-Nordgren E, Göransson P. Optimising open porous foam for acoustical and vibrational performance. Journal of Sound and Vibration. 2010; **329**:753-767. DOI: 10.1016/j.jsv.2009.10.009

[5] Naldzhiev D, Mumovic D, Strlic M. Polyurethane insulation and household products—A systematic review of their impact on indoor environmental quality. Building and Environment. 2020;**169**:1-18. Article ID 106559. DOI: 10.1016/j.buildenv.2019.106559

[6] Andreola F, Barbieri L, Lancellotti I, Leonelli C, Manfredini T. Recycling of industrial wastes in ceramic manufacturing: State of art and glass case studies. Ceramics International. 2016;**42**:13333-13338. DOI: 10.1016/j.ceramint.2016.05.205

[7] Benkreira H, Khan A, Horoshenkov KV. Sustainable acoustic and thermal insulation materials from elastomeric waste residues. Chemical Engineering Science. 2011;**66**:4157-4171. DOI: 10.1016/j.ces.2011.05.047

[8] Khan A, Mohamed M, Al Halo N, Benkreira H. Acoustical properties of novel sound absorbers made from recycled granulates. Applied Acoustics. 2017;**127**:80-88. DOI: 10.1016/j.apacoust.2017.05.035

[9] Bernardo E, Cedro R, Florean M, Hreglich S. Reutilization and stabilization of wastes by the production of glass foams. Ceramics International. 2007;**33**:963-968. DOI: 10.1016/j.ceramint.2006.02.010

[10] Bernardo E, Scarinci G, Bertuzzi P, Ercole P, Ramon L. Recycling of waste glasses into partially crystallized glass foams. Journal of Porous Materials. 2010;**17**:359-365. DOI: 10.1007/s10934-009-9286-3

[11] Chinnam RK, Francis AA, Will J, Bernardo E, Boccaccini AR. Review. Functional glasses and glass-ceramics derived from iron rich waste and combination of industrial residues. Journal of Non-Crystalline Solids. 2013; **365**:63-74. DOI: 10.1016/j.jnoncrysol.2012.12.006

[12] Liu Y, Farnsworth M, Tiwari A. A review of optimisation techniques used in the composite recycling area: State-of-the-art and steps towards a research agenda. Journal of Cleaner Production. 2017;**140**:1775-1781. DOI: 10.1016/j.jclepro.2016.08.038

[13] Overcash M, Twomey J, Asmatulu E, Vozzola E, Griffing E. Thermoset composite recycling—Driving forces, development, and evolution of new opportunities. Journal of Composite Materials. 2018;**52**: 1033-1043. DOI: 10.1177/0021998317720000

[14] Caniato M, Travan A. Method for recycling waste material. 2016. EP 3216825, filed 11 March 2016 and issued 28 August 2019

[15] Caniato M, D'Amore GKO, Kaspar J, Gasparella A. Sound absorption

performance of sustainable foam materials: Application of analytical and numerical tools for the optimization of forecasting models. Applied Acoustics. 2020;**161**:107166. DOI: 10.1016/j.apacoust.2019.107166

[16] Allard JF, Atalla N. Propagation of Sound in Porous Media: Modelling Sound Absorbing Materials. 2nd ed. Chichester, UK: John Wiley & Sons; 2009

[17] Allard J, Champoux Y. New empirical equations for sound propagation in rigid frame fibrous materials. The Journal of the Acoustical Society of America. 1992;**91**:3346-3353. DOI: 10.1121/1.402824

[18] Öchsner A, Murch GE, de Lemos MJS, editors. Cellular and Porous Materials: Thermal Properties Simulation and Prediction. Weinheim: Wiley-VCH Verlag GmbH & Co. KGaA; 2008. p. 422. DOI: 10.1002/9783527621408

[19] Placido E, Arduini-Schuster MC, Kuhn J. Thermal properties predictive model for insulating foams. Infrared Physics & Technology. 2005;**46**:219-231. DOI: 10.1016/j.infrared.2004.04.001

[20] Randrianalisoa J, Baillis D. Thermal conductive and radiative properties of solid foams: Traditional and recent advanced modelling approaches. Comptes Rendus Physique. 2014;**15**:683-695. DOI: 10.1016/j.crhy.2014.09.002

[21] Tseng C-J, Yamaguchi M, Ohmori T. Thermal conductivity of polyurethane foams from room temperature to 20 K. Cryogenics. 1997;**37**:305-312. DOI: 10.1016/S0011-2275(97)00023-4

[22] Cunsolo S, Coquard R, Baillis D, Chiu WKS, Bianco N. Radiative properties of irregular open cell solid foams. International Journal of Thermal Sciences. 2017;**117**:77-89. DOI: 10.1016/j.ijthermalsci.2017.03.007

[23] Mendes MAA, Ray S, Trimis D. A simple and efficient method for the evaluation of effective thermal conductivity of open-cell foam-like structures. International Journal of Heat and Mass Transfer. 2013;**66**:412-422. DOI: 10.1016/j.ijheatmasstransfer.2013.07.032

[24] Klett JW, McMillan AD, Gallego NC, Walls CA. The role of structure on the thermal properties of graphitic foams. Journal of Materials Science. 2004;**39**:3659-3676. DOI: 10.1023/B:JMSC.0000030719.80262.f8

[25] Pande RN, Kumar V, Chaudhary DR. Thermal conduction in a homogeneous two-phase system. Pramana. 1984;**22**:63-70. DOI: 10.1007/BF02875588

[26] Jagjiwanram R. Singh, effective thermal conductivity of two-phase systems with cylindrical inclusions. Indian Journal of Pure and Applied Physics. 2004;**42**:600-609

[27] Russell HW. Principles of heat flow in porous insulators. Journal of the American Ceramic Society. 1935;**18**:1-5. DOI: 10.1111/j.1151-2916.1935.tb19340.x

[28] Belova IV, Murch GE. Thermal Properties of Composite Materials and Porous Media: Lattice-Based Monte Carlo Approaches. 2008. 10.1002/9783527621408.ch3

[29] Yüksel N. The review of some commonly used methods and techniques to measure the thermal conductivity of insulation materials. In: Almusaed A, editor. Insulation Materials in Context of Sustainability. London: IntechOpen; 2016. p. 114-140. DOI: 10.5772/64157

[30] Cao L, Fu Q, Si Y, Ding B, Yu J. Porous materials for sound absorption.

Composites Communications. 2018;**10**: 25-35. DOI: 10.1016/j.coco.2018.05.001

[31] Bujoreanu C, Nedeff F, Benchea M, Agop M. Experimental and theoretical considerations on sound absorption performance of waste materials including the effect of backing plates. Applied Acoustics. 2017;**119**:88-93. DOI: 10.1016/j.apacoust.2016.12.010

[32] Peng L, Song B, Wang J, Wang D. Mechanic and acoustic properties of the sound-absorbing material made from natural fiber and polyester. Advances in Materials Science and Engineering. 2015; **2015**:1-5. Article ID 274913. DOI: 10.1155/ 2015/274913

[33] Rahimabady M, Statharas EC, Yao K, Sharifzadeh Mirshekarloo M, Chen S, Tay FEH. Hybrid local piezoelectric and conductive functions for high performance airborne sound absorption. Applied Physics Letters. 2017;**111**:1-4. Article ID 241601. DOI: 10.1063/1.5010743

[34] Biot MA. Theory of propagation of elastic waves in a fluid-saturated porous solid II. Higher frequency range. The Journal of the Acoustical Society of America. 1956;**28**:179-191. DOI: 10.1121/ 1.1908241

[35] Biot MA. Theory of propagation of elastic waves in a fluid-saturated porous solid. I. Low-frequency range. The Journal of the Acoustical Society of America. 1956;**28**:168-178. DOI: 10.1121/ 1.1908239

[36] Biot MA. Mechanics of deformation and acoustic propagation in porous media. Journal of Applied Physics. 1962; **33**:1482-1498. DOI: 10.1063/1.1728759

[37] Egab L, Wang X, Fard M. Acoustical characterisation of porous sound absorbing materials: A review. International Journal of Vehicle Noise and Vibration. 2014;**10**:129-149. DOI: 10.1504/IJVNV.2014.059634

[38] Horoshenkov KV. A review of acoustical methods for porous material characterisation. International Journal of Acoustics and Vibration. 2017;**22**: 92-103. DOI: 10.20855/ijav.2017.22.1455

[39] Oliva D, Hongisto V. Sound absorption of porous materials—Accuracy of prediction methods. Applied Acoustics. 2013;**74**: 1473-1479. DOI: 10.1016/j.apacoust. 2013.06.004

[40] Kidner MRF, Hansen CH. A comparison and review of theories of the acoustics of porous materials. International Journal of Acoustics and Vibrations. 2008;**13**:112-119

[41] Perrot C, Chevillotte F, Panneton R. Bottom-up approach for microstructure optimization of sound absorbing materials. The Journal of the Acoustical Society of America. 2008;**124**:940-948. DOI: 10.1121/1.2945115

[42] Perrot C, Hoang MT, Chevillotte F. An overview of microstructural approaches for modelling and improving sound proofing properties of cellular foams: Developments and prospects, SAE Technical Papers. June 2018. pp. 1-8. DOI: 10.4271/2018-01-1564

[43] Doutres O, Atalla N, Dong K. Effect of the microstructure closed pore content on the acoustic behavior of polyurethane foams. Journal of Applied Physics. 2011;**110**:1-11. Article ID 064901. DOI: 10.1063/1.3631021

[44] Abdessalam H, Abbès B, Abbès F, Li Y, Guo Y-Q. Prediction of acoustic properties of polyurethane foams from the macroscopic numerical simulation of foaming process. Applied Acoustics. 2017;**120**:129-136. DOI: 10.1016/j. apacoust.2017.01.021

[45] Johnson DL, Koplik J, Dashen R. Theory of dynamic permeability and tortuosity in fluid-saturated porous media. Journal of Fluid Mechanics. 1987;

176:379. DOI: 10.1017/S002211 2087000727

[46] Champoux Y, Allard J. Dynamic tortuosity and bulk modulus in air-saturated porous media. Journal of Applied Physics. 1991;**70**:1975-1979. DOI: 10.1063/1.349482

[47] Panneton R. Comments on the limp frame equivalent fluid model for porous media. The Journal of the Acoustical Society of America. 2007;**122**:EL217-EL222. DOI: 10.1121/1.2800895

[48] Kino N. Further investigations of empirical improvements to the Johnson–Champoux–Allard model. Applied Acoustics. 2015;**96**:153-170. DOI: 10.1016/j.apacoust.2015.03.024

[49] Fotsing ER, Dubourg A, Ross A, Mardjono J. Acoustic properties of periodic micro-structures obtained by additive manufacturing. Applied Acoustics. 2019;**148**:322-331. DOI: 10.1016/j.apacoust.2018.12.030

[50] Tarnow V. Airflow resistivity of models of fibrous acoustic materials. The Journal of the Acoustical Society of America. 1996;**100**:3706-3713. DOI: 10.1121/1.417233

[51] Deckers E, Jonckheere S, Vandepitte D, Desmet W. Modelling techniques for vibro-acoustic dynamics of poroelastic materials. Archives of Computational Methods in Engineering. 2015;**22**:183-236. DOI: 10.1007/s11831-014-9121-0

[52] Bonfiglio P, Pompoli F. Inversion problems for determining physical parameters of porous materials: Overview and comparison between different methods. Acta Acustica united with Acustica. 2013;**99**:341-351. DOI: 10.3813/AAA.918616

[53] Bonfiglio P, Pompoli F, Shravage P. Quasistatic evaluation of mechanical properties of poroelastic materials: Static and dynamic strain dependence and in vacuum tests. The Journal of the Acoustical Society of America. 2008; **123**:3037-3037. DOI: 10.1121/1.2932705

[54] Scheffler M, Colombo P, editors. Cellular Ceramics: Structure, Manufacturing, Properties and Applications. Weinheim: Wiley-VCH; 2005

[55] Studart AR, Gonzenbach UT, Tervoort E, Gauckler LJ. Processing routes to macroporous ceramics: A review. Journal of the American Ceramic Society. 2006;**89**:1771-1789. DOI: 10.1111/j.1551-2916.2006.01044.x

[56] Colombo P, Vakifahmetoglu C, Costacurta S. Fabrication of ceramic components with hierarchical porosity. Journal of Materials Science. 2010;**45**: 5425-5455. DOI: 10.1007/s10853-010-4708-9

[57] Fernandes HR, Tulyaganov DU, Ferreira JMF. Preparation and characterization of foams from sheet glass and fly ash using carbonates as foaming agents. Ceramics International. 2009;**35**:229-235. DOI: 10.1016/j.ceramint.2007.10.019

[58] Asdrubali F, Schiavoni S, Horoshenkov KV. A review of sustainable materials for acoustic applications. Building Acoustics. 2012;**19**:283-312. DOI: 10.1260/1351-010X.19.4.283

[59] Asdrubali F, D'Alessandro F, Schiavoni S. A review of unconventional sustainable building insulation materials. Sustainable Materials and Technologies. 2015;**4**:1-17. DOI: 10.1016/j.susmat.2015.05.002

[60] Winterton N. Twelve more green chemistry principles. Green Chemistry. 2001;**3**:G73-G75. DOI: 10.1039/b110187k

[61] Lee KY, Mooney DJ. Alginate: Properties and biomedical applications.

Progress in Polymer Science. 2012;**37**: 106-126. DOI: 10.1016/j.progpolymsci.2011.06.003

[62] Goh CH, Heng PWS, Chan LW. Alginates as a useful natural polymer for microencapsulation and therapeutic applications. Carbohydrate Polymers. 2012;**88**:1-12. DOI: 10.1016/j.carbpol.2011.11.012

[63] Vincent T, Dumazert L, Dufourg L, Cucherat C, Sonnier R, Guibal E. New alginate foams: Box-Behnken design of their manufacturing; fire retardant and thermal insulating properties: Article. Journal of Applied Polymer Science. 2018;**135**:45868. DOI: 10.1002/app.45868

[64] Galiano F, Briceño K, Marino T, Molino A, Christensen KV, Figoli A. Advances in biopolymer-based membrane preparation and applications. Journal of Membrane Science. 2018;**564**:562-586. DOI: 10.1016/j.memsci.2018.07.059

[65] Shaari N, Kamarudin SK. Chitosan and alginate types of bio-membrane in fuel cell application: An overview. Journal of Power Sources. 2015;**289**: 71-80. DOI: 10.1016/j.jpowsour.2015.04.027

[66] Rechberger F, Niederberger M. Synthesis of aerogels: From molecular routes to 3-dimensional nanoparticle assembly. Nanoscale Horizons. 2017;**2**:6-30. DOI: 10.1039/c6nh00077k

[67] Plieva FM, Kirsebom H, Mattiasson B. Preparation of macroporous cryostructurated gel monoliths, their characterization and main applications. Journal of Separation Science. 2011;**34**:2164-2172. DOI: 10.1002/jssc.201100199

[68] Mahler W, Bechtold MF. Freeze-formed silica fibres. Nature. 1980;**285**: 27-28. DOI: 10.1038/285027a0

[69] Zhang H, Cooper AI. Aligned porous structures by directional freezing. Advanced Materials. 2007;**19**:1529-1533. DOI: 10.1002/adma.200700154

[70] Porrelli D, Travan A, Turco G, Marsich E, Borgogna M, Paoletti S, et al. Alginate-hydroxyapatite bone scaffolds with isotropic or anisotropic pore structure: Material properties and biological behavior. Macromolecular Materials and Engineering. 2015;**300**: 989-1000. DOI: 10.1002/mame.201500055

[71] D'Amore GKO. Studio e caratterizzazione mediante analisi FEM di una schiuma innovativa per l'isolamento termico ed acustico di porte tagliafuoco ad uso navale (Study and characterization by FEM analysis of an innovative foam for thermal and acoustic insulation of naval fire-doors), [MS thesis]. University of Trieste; 2018

[72] Travan A, Scognamiglio F, Borgogna M, Marsich E, Donati I, Tarusha L, et al. Hyaluronan delivery by polymer demixing in polysaccharide-based hydrogels and membranes for biomedical applications. Carbohydrate Polymers. 2016;**150**:408-418. DOI: 10.1016/j.carbpol.2016.03.088

[73] Olivas GI, Barbosa-Cánovas GV. Alginate–calcium films: Water vapor permeability and mechanical properties as affected by plasticizer and relative humidity. LWT—Food Science and Technology. 2008;**41**:359-366. DOI: 10.1016/j.lwt.2007.02.015

[74] Scognamiglio F, Travan A, Borgogna M, Donati I, Marsich E, Bosmans JWAM, et al. Enhanced bioadhesivity of dopamine-functionalized polysaccharidic membranes for general surgery applications. Acta Biomaterialia. 2016;**44**:232-242. DOI: 10.1016/j.actbio.2016.08.017

[75] Scognamiglio F, Travan A, Donati I, Borgogna M, Marsich E, Andersen T,

et al. H_2O_2 causes improved adhesion between a polysaccharide-based membrane and intestinal serosa. Colloid and Interface Science Communications. 2016;**15**:5-8. DOI: 10.1016/j.colcom.2016.11.002

[76] Verdejo R, Stämpfli R, Alvarez-Lainez M, Mourad S, Rodriguez-Perez MA, Brühwiler PA, et al. Enhanced acoustic damping in flexible polyurethane foams filled with carbon nanotubes. Composites Science and Technology. 2009;**69**:1564-1569. DOI: 10.1016/j.compscitech.2008.07.003

[77] Oliveux G, Dandy LO, Leeke GA. Current status of recycling of fibre reinforced polymers: Review of technologies, reuse and resulting properties. Progress in Materials Science. 2015;**72**:61-99. DOI: 10.1016/j.pmatsci.2015.01.004

[78] Zhang H, Hussain I, Brust M, Butler MF, Rannard SP, Cooper AI. Aligned two- and three-dimensional structures by directional freezing of polymers and nanoparticles. Nature Materials. 2005;**4**:787-793. DOI: 10.1038/nmat1487

[79] Li WL, Lu K, Walz JY. Freeze casting of porous materials: Review of critical factors in microstructure evolution. International Materials Review. 2012;**57**:37-60. DOI: 10.1179/1743280411Y.0000000011

[80] Deville S. Freeze-casting of porous ceramics: A review of current achievements and issues. Advanced Engineering Materials. 2008;**10**:155-169. DOI: 10.1002/adem.200700270

[81] Deville S, Maire E, Lasalle A, Bogner A, Gauthier C, Leloup J, et al. Influence of particle size on ice nucleation and growth during the ice-templating process. Journal of the American Ceramic Society. 2010;**93**:2507-2510. DOI: 10.1111/j.1551-2916.2010.03840.x

[82] Deville S, Saiz E, Tomsia AP. Ice-templated porous alumina structures. Acta Materialia. 2007;**55**:1965-1974. DOI: 10.1016/j.actamat.2006.11.003

[83] Archie GE. The electrical resistivity as an aid in core analysis interpretation. Transactions of the American Institute of Mining and Metallurgical Engineers. 1942;**146**:54-62

[84] Allard J, Atalla N. Propagation of Sound in Porous Media Modelling Sound Absorbing Materials. 2nd ed. London: Wiley; 2009

[85] Yang XH, Ren SW, Wang WB, Liu X, Xin FX, Lu TJ. A simplistic unit cell model for sound absorption of cellular foams with fully/semi-open cells. Composites Science and Technology. 2015;**118**:276-283. DOI: 10.1016/j.compscitech.2015.09.009

Chapter 3

Turbulent Heat Transfer Analysis of Silicon Carbide Ceramic Foam as a Solar Volumetric Receiver

Chen Yang, Huijin Xu and Akira Nakayama

Abstract

A volumetric solar receiver receives the concentrated radiation generated by a large number of heliostats. Turbulent heat transfer occurs from the solid matrix to the air as it passes through the porous receiver. Such combined heat transfer within the receiver, including radiation, convection and conduction, is studied using a local thermal non-equilibrium model. Both the Rosseland approximation and the P1 model are applied to consider the radiative heat transfer through the solar receiver. Furthermore, the low Mach approximation is exploited to investigate the compressible flow through the receiver. Analytic solutions are obtained for the developments of air and ceramic temperatures as well as the pressure along the flow direction. Since the corresponding fluid and solid temperature variations generated under the Rosseland approximation agree fairly well with those based on the P1 model, the Rosseland approximation is used for further analysis. The results indicate that the pore diameter must be larger than its critical value to obtain high receiver efficiency. Moreover, it has been found that optimal pore diameter exists for achieving the maximum receiver efficiency under the equal pumping power. The solutions provide effective guidance for a novel volumetric solar receiver design of silicon carbide ceramic foam.

Keywords: turbulent heat transfer, thermal non-equilibrium, Rosseland approximation, P1 model, volumetric solar receiver, porous media, ceramic foam

1. Introduction

A solar volumetric receiver is required to have the resistance to temperature as high as 1000 degree Celsius, high porosity for sufficiently large extinction volume such that the concentrated solar radiation penetrates through the receiver, high cell density to achieve large specific surface area and sufficiently high effective thermal conductivity to avoid possible thermal spots. Extruded monoliths with parallel channels (i.e. honeycomb structure) are being used in some solar power plants in Europe, including the solar power tower plant of 1.5 MW built in 2009, in Julich in Germany [1, 2]. However, in such conventional receivers, both thermal spots [3] and flow instabilities [4] have been often reported. In the monolith receiver, locally high solar flux leads to a low mass flow with high temperature, whereas locally low solar flux leads to a high mass flow with low temperature. This causes the absorber material to exceed the design temperature locally, which then leads to its destruction although the average temperature is comparatively low. These difficulties encountered in the receiver must be overcome to run the power plant safely.

In consideration of these requirements, ceramic foams have come to draw attention as a possible candidate to replace the conventional extruded monoliths with parallel channels. Many researchers including Becker et al. [4], Fend et al. [5] and Bai [6] focused on porous ceramic foams as a promising absorber material. Recently, Sano et al. [7] carried out a local non-thermal equilibrium analysis to investigate the receiver efficiency under the equal pumping power. For the first time, the complete set of analytical solutions based on the two-energy equation model of porous media was presented, so as to fully account for the combined effects of tortuosity; thermal dispersion and compressibility on the convective, conductive and radiative heat transfer within a ceramic foam receiver. In their analysis, however, the Rosseland approximation was applied to account for the radiative heat transfer through the solar receiver. It is well known that the Rosseland approximation ceases to be valid near boundaries. Although no wall boundaries exist for the case of the one-dimensional analysis of the solar volumetric receiver, the validity of applying the Rosseland approximation near the inlet boundary of the receiver has not been investigated yet. Furthermore, the effects of turbulence mixing on the heat transfer were not considered.

In this study, the validity of the Rosseland approximation [7] will be examined by comparing the results based on the Rosseland approximation and the results obtained from solving the irradiation transport equation based on the P1 model. The set of the equations will be reduced to a fifth-order ordinary differential equation for the air temperature. Once the air temperature distribution is determined, the pressure distribution along the flow direction can readily be estimated from the momentum equation with the low Mach approximation. Thus, the receiver efficiency, namely, the ratio of the air enthalpy flux increase to the concentrated solar heat flux, can be compared under the equal pumping power, so as to investigate the optimal operating conditions. Some analytical and numerical investigations [3–8] have been reported elsewhere. However, none of them appeared to elucidate well the combined effects of turbulence, compressibility, radiation, convection and conduction within the volumetric receiver on the developments of air and ceramic temperatures as well as the pressure along the flow direction. This study appears to be the first to provide the complete set of analytical solutions based on the two-energy equation model of porous media [9], fully accounting for the combined effects of turbulence, tortuosity, thermal dispersion, compressibility and radiative heat transfer within a ceramic foam receiver.

2. Volume averaged governing equations

As illustrated in **Figure 1**, the structure of silicon carbide ceramic foam volumetric receiver may be treated as homogeneous porous medium. Since the dependence of the Darcian velocity on the transverse direction can only be observed in a small region very close to the walls of the passage, we may neglect the boundary effects (i.e. Brinkman term).

Based on a theoretical derivation of Darcy's law, Neuman [10] pointed out that the application of Darcy's law to compressible fluids is justified as long as Knudsen numbers are sufficiently small to ensure the no-slip conditions at the solid–gas interface. This is usually the case for the volumetric receivers. Thus, allowing the density to vary through the receiver, the following Forchheimer extended Darcy law should hold:

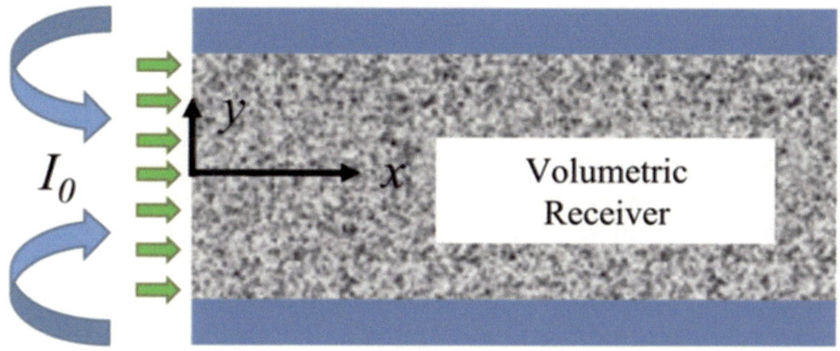

Figure 1.
Volumetric receiver.

$$-\frac{\partial \langle p \rangle^f}{\partial x_i} = \frac{\langle \mu \rangle^f}{K} \langle u_i \rangle + b \langle \rho \rangle^f \sqrt{\langle u_j \rangle \langle u_j \rangle} \langle u_i \rangle \tag{1}$$

where K and b are the permeability and the inertial coefficients, respectively. Furthermore, by virtue of the volume averaging procedure [11–13], the microscopic energy equations of the compressible fluid flow phase and the solid phase may be integrated over an elemental control volume V, so as to derive the corresponding macroscopic energy equations. Since the porous medium is considered to be homogeneous, the integration of the two distinct energy equations gives:

For the air:

$$\begin{aligned} &\varepsilon \frac{\partial}{\partial t} \left\langle \rho_f \left(h_{stag} - \frac{p}{\rho} \right) \right\rangle^f + \varepsilon \frac{\partial}{\partial x_j} \langle \rho \rangle^f \langle u_j \rangle^f \langle h_{st} \rangle^f \\ &= \frac{\partial}{\partial x_j} \left(\varepsilon \langle k_f \rangle^f \frac{\partial \langle T \rangle^f}{\partial x_j} + \frac{1}{V} \int_{A_{int}} k_f T n_j dA - \varepsilon \langle \rho \rangle^f \left\langle \tilde{h}_{stag} \tilde{u}_j \right\rangle^f + \varepsilon \langle u_i \tau_{ij} \rangle^f \right) \\ &+ \frac{1}{V} \int_{A_{int}} k_f \frac{\partial T}{\partial x_j} n_j dA \end{aligned} \tag{2}$$

For the solid matrix:

$$(1-\varepsilon)\rho_s c_s \frac{\partial \langle T \rangle^s}{\partial t} = \frac{\partial}{\partial x_j} \left((1-\varepsilon) k_s \frac{\partial \langle T \rangle^s}{\partial x_j} - \frac{k_s}{V} \int_{A_{int}} T n_j dA \text{-} q_R \right) - \frac{1}{V} \int_{A_{int}} k_f \frac{\partial T}{\partial x_j} n_j dA \tag{3}$$

where the intrinsic volume average of a certain local variable ϕ in the fluid phase and solid matrix phase can be defined as.

$$\langle \phi \rangle^f \equiv \frac{1}{V_f} \int_{V_f} \phi dV, \langle \phi \rangle^m \equiv \frac{1}{V_m} \int_{V_m} \phi dV \tag{4}$$

Note that subscripts f and m refer to the fluid phase and solid matrix phase, respectively. The decomposition of the local variable ϕ can be expressed in terms of its intrinsic average and the spatial deviation from it:

$$\phi = \langle \phi \rangle^f + \tilde{\phi} \tag{5}$$

Moreover, q_R is the radiative heat flux, A_{int} is the interfacial surface area between the fluid and solid matrix phases, while n_j is the normal unit vector from the fluid phase to the solid matrix phase.

In order to simplify the foregoing set of the equations, the low Mach approximation is applied due to the relatively low Mach number when the air flows through a porous medium. Thus, the dynamic pressure change is sufficiently small as compared to the absolute pressure prevailing over the system, such that the stagnant enthalpy is approximated by $h_{stag} = h + u_k u_k / 2 \cong h$. Combining the foregoing two energy equations namely Eqs. (2) and (3), and, then, noting the continuity of temperature and heat flux at the interface, we obtain the one-equation model for the steady state as follows:

$$\varepsilon \frac{\partial}{\partial x_j} \left\langle \rho_f \right\rangle^f \langle u_j \rangle^f \langle h \rangle^f = \frac{\partial}{\partial x_j} \left(\varepsilon \langle k_f \rangle^f \frac{\partial \langle T \rangle^f}{\partial x_j} + (1-\varepsilon) k_s \frac{\partial \langle T \rangle^s}{\partial x_j} + \frac{1}{V} \int_{A_{\text{int}}} (k_f - k_s) T n_j dA - \varepsilon \left\langle \rho_f \right\rangle^f \left\langle \tilde{h} \tilde{u}_j \right\rangle^f - q_R \right) \tag{6}$$

For the time being, let us assume $\partial \langle T \rangle^f / \partial x_j \cong \partial \langle T \rangle^s / \partial x_j \cong \partial \langle T \rangle / \partial x_j$ (this assumption will be relaxed shortly). Then, the equation reduces to

$$\varepsilon \frac{\partial}{\partial x_j} \left\langle \rho_f \right\rangle^f \langle u_j \rangle^f \langle h \rangle^f = \frac{\partial}{\partial x_j} \left(\left(\varepsilon \langle k_f \rangle^f + (1-\varepsilon) k_s \right) \frac{\partial \langle T \rangle}{\partial x_j} + \frac{1}{V} \int_{A_{\text{int}}} (k_f - k_s) T n_j dA - \varepsilon \left\langle \rho_f \right\rangle^f \left\langle \tilde{h} \tilde{u}_j \right\rangle^f - q_R \right) \tag{7}$$

where

$$\langle \phi \rangle \equiv \frac{1}{V} \int_V \phi dV \tag{8}$$

is the Darcian average of the variable ϕ such that $\langle u_j \rangle = \varepsilon \langle u_j \rangle^f$ is the Darcian velocity vector. From the foregoing equation, that is, Eq. (6), the macroscopic heat flux vector $q_i = \left(q_x, q_y, q_z \right)$ and its corresponding stagnant thermal conductivity k_{stag} may be defined as follows:

$$\begin{aligned} q_i &= -k_{stag} \frac{\partial \langle T \rangle}{\partial x_i} + q_R + \varepsilon \left\langle \rho_f \right\rangle^f \left\langle \tilde{h} \tilde{u}_i \right\rangle^f \\ &= -\left(\varepsilon \langle k_f \rangle^f + (1-\varepsilon) k_s \right) \frac{\partial \langle T \rangle}{\partial x_i} - \frac{1}{V} \int_{A_{\text{int}}} (k_f - k_s) T n_i dA + q_R + \varepsilon \left\langle \rho_f \right\rangle^f \left\langle \tilde{h} \tilde{u}_i \right\rangle^f \end{aligned} \tag{9}$$

or

$$k_{stag}\frac{\partial\langle T\rangle}{\partial x_i} = \left(\varepsilon\langle k_f\rangle^f + (1-\varepsilon)k_s\right)\frac{\partial\langle T\rangle}{\partial x_i} + \frac{1}{V}\int_{A_{\rm int}}(k_f - k_s)Tn_i dA \tag{10}$$

The term $\varepsilon\left\langle\rho_f\right\rangle^f\left\langle\tilde{h}\tilde{u}_i\right\rangle^f$ in Eq. (9) describes the thermal dispersion heat flux vector, which serves an additional heat flux resulting from the hydrodynamic mixing of fluid particles passing through pores. On the other hand, the second term on the right-hand side term in Eq. (10) is associated with the surface integral, and it describes the effects of the tortuosity on the macroscopic heat flux, which adjusts the level of the stagnant thermal conductivity from its upper bound $\left(\varepsilon k_f + (1-\varepsilon)k_s\right)$ to a correct one. Yang and Nakayama [9] introduced the effective porosity ε^*, which is defined as

$$\varepsilon^* = \frac{k_s - k_{stag}}{k_s - k_f} = \varepsilon + \frac{\varepsilon k_f + (1-\varepsilon)k_s - k_{stag}}{k_s - k_f} \tag{11}$$

such that

$$(\varepsilon^* - \varepsilon)\frac{\partial\langle T\rangle}{\partial x_i} = \frac{1}{V}\int_{A_{\rm int}} Tn_i dA \tag{12}$$

Using the effective porosity ε^* and the equation of state $\langle p\rangle^f = \left\langle\rho_f\right\rangle^f R\langle T\rangle^f$ and $\langle h\rangle^f = c_p\langle T\rangle^f$, the volume average energy equations Eqs. (2) and (3) may be concisely rewritten for the steady state for air as:

$$\varepsilon c_p\frac{\partial}{\partial x_j}\left\langle\rho_f\right\rangle^f\langle u_j\rangle^f\langle T\rangle^f = \frac{\partial}{\partial x_j}\left(\varepsilon^*\langle k_f\rangle^f\frac{\partial\langle T\rangle^f}{\partial x_j} + \varepsilon k_{dis_{jk}}\frac{\partial\langle T\rangle^f}{\partial x_k}\right) - h_v\left(\langle T\rangle^f - \langle T\rangle^s\right) \tag{13}$$

for the solid matrix phase as:

$$\frac{\partial}{\partial x_j}\left((1-\varepsilon^*)k_s\frac{\partial\langle T\rangle^s}{\partial x_j} + q_R\right) - h_v\left(\langle T\rangle^s - \langle T\rangle^f\right) = 0 \tag{14}$$

Note that the assumption of equal temperature gradients, $\partial\langle T\rangle^f/\partial x_j \cong \partial\langle T\rangle^s/\partial x_j \cong \partial\langle T\rangle/\partial x_j$, has been discarded. This practice has been proven to be quite effective in a series of computations (e.g. [8, 9]). According to the gradient diffusion hypothesis [14], the thermal dispersion term is usually expressed as:

$$\left\langle\rho_f\right\rangle^f\left\langle\tilde{h}\tilde{u}_j\right\rangle^f = \left\langle\rho_f\right\rangle^f c_p\left\langle\tilde{T}\tilde{u}_j\right\rangle^f = -k_{disjk}\frac{\partial\langle T\rangle^f}{\partial x_k} \tag{15}$$

while the interfacial heat transfer between the solid and fluid phases is modeled using Newton's cooling law:

$$\frac{1}{V}\int_{A_{\rm int}} k_f\frac{\partial T}{\partial x_j}n_j dA = h_v\left(\langle T\rangle^s - \langle T\rangle^f\right) \tag{16}$$

where h_v is the volumetric heat transfer coefficient. The Maxwell approximations may be used for the dynamic viscosity and thermal conductivity of the air:

$$\mu\left(\langle T\rangle^f\right) = \mu_0\left(\frac{\langle T\rangle^f}{\langle T\rangle_0^f}\right)^n = 1.8\times10^{-5}\left(\frac{\langle T\rangle^f}{300K}\right)^{0.7}\ [\mathrm{Pa\cdot s}] \tag{17}$$

and

$$k_f\left(\langle T\rangle^f\right) = k_0\left(\frac{\langle T\rangle^f}{\langle T\rangle_0^f}\right)^n = 0.025\left(\frac{\langle T\rangle^f}{300K}\right)^{0.7}\ [\mathrm{W/mK}] \tag{18}$$

where the exponent n is 0.7 according to [4]. The specific heat capacity of the air $c_p = 1000[\mathrm{J/kgK}]$ and the Prandtl number $Pr = 1.8\times10^{-5}\times1000/0.025 = 0.72$ are assumed to be constant.

3. One-dimensional analysis for volumetric receiver

In this section, we perform one-dimensional analysis to obtain analytic solutions for convective-radiative heat transfer in volume receiver. Prior to that, the radiative heat flux q_R needs to be determined in advance. In the literature, there are two models, namely, the Rosseland approximation and the P1 model.

3.1 Analysis based on the Rosseland approximation

In the Rosseland approximation, the radiative heat flux is given by

$$q_R = -\frac{16\sigma}{3\beta}\left(\langle T\rangle^s\right)^3\frac{\partial\langle T\rangle^s}{\partial x_j} \tag{19}$$

where $\sigma = 5.67\times10^{-8}\left[\mathrm{W/m^2K^4}\right]$ is the Stephan-Boltzmann constant while β is the mean extinction coefficient.

As schematically shown in **Figure 1**, the air is flowing through a passage of length L at the rate of the mass flux $G = \left\langle\rho_f\right\rangle^f\langle u\rangle$. Under the low Mach number approximation, namely, $\left\langle\rho_f\right\rangle^f \propto 1/\langle T\rangle^f$, the macroscopic governing equations Eqs. (1), (13) and (14) can be simplified to be a one-dimensional set of equations as follows:

$$-\frac{d\langle p\rangle^f}{dx} = \frac{\langle\mu\rangle^f}{K}\frac{G}{\left\langle\rho_f\right\rangle^f} + b\frac{G^2}{\left\langle\rho_f\right\rangle^f} = \frac{R}{\langle p\rangle^f}\left(\frac{\langle\mu\rangle^f}{K}G + bG^2\right)\langle T\rangle^f \tag{20}$$

$$c_pG\frac{d\langle T\rangle^f}{dx} = \frac{d}{dx}\left(\varepsilon^*\left\langle k_f\right\rangle^f + \varepsilon k_{disxx}\right)\frac{d\langle T\rangle^f}{dx} - h_v\left(\langle T\rangle^f - \langle T\rangle^s\right) \tag{21}$$

$$\frac{d}{dx}\left((1-\varepsilon^*)k_s + \frac{16\sigma}{3\beta}\left(\langle T\rangle^s\right)^3\right)\frac{d\langle T\rangle^s}{dx} - h_v\left(\langle T\rangle^s - \langle T\rangle^f\right) = 0 \tag{22}$$

According to Calmidi and Mahajan [15, 16], Dukhan [17], Kuwahara et al. [18] and Yang et al. [19, 20], the permeability and inertial coefficient of foams are given by

$$K = 0.00073(1-\varepsilon)^{-0.224}\left(\frac{1.18}{1-e^{-(1-\varepsilon)/0.04}}\sqrt{\frac{1-\varepsilon}{3\pi}}\right)^{-1.11}{d_m}^2 \tag{23}$$

and

$$b = \frac{12(1-\varepsilon)}{d_m} \tag{24}$$

respectively, where d_m is the pore diameter of foam. The longitudinal dispersion coefficient is roughly about 20 times more than the transverse one. Thus, following Calmidi and Mahajan [16], we may evaluate the longitudinal dispersion coefficient using the following expression:

$$\varepsilon k_{dis_{xx}} = 1.2c_p G\sqrt{K} \tag{25}$$

With respect to the stagnant thermal conductivity and the volumetric heat transfer coefficient for foams, Calmidi and Mahajan [15, 16] empirically provided the following correlations:

$$k_{stag} = \varepsilon k_f + 0.19(1-\varepsilon)^{0.763}k_s \tag{26}$$

$$h_v = 8.72(1-\varepsilon)^{1/4}\left(\frac{1-e^{-(1-\varepsilon)/0.04}}{\varepsilon}\right)^{1/2}\left(\frac{Gd_m}{\langle\mu\rangle^f}\right)^{1/2}Pr^{0.37}\frac{k_f}{{d_m}^2} \tag{27}$$

Kamiuto et al. [21] experimentally affirmed that the Rosseland model is quite effective. Therefore, it can be deduced that the Rosseland model is also applicable for the present case of silicon carbide ceramic foam. Based on the measurements made on cordierite ceramic foams by Kamiuto et al., the mean extinction coefficient β is calculated by the following correlation:

$$\beta = 8(1-\varepsilon)/d_m \tag{28}$$

For a given mass flux $G = \left\langle\rho_f\right\rangle^f\langle u\rangle$, the foregoing three equations along with the equation of state may be solved for the four unknowns, namely, $\langle T\rangle^f$,$\langle T\rangle^s$, $\langle p\rangle^f$ and $\left\langle\rho_f\right\rangle^f$. The boundary conditions are given as follows:

$x = 0$ (inlet):

$$\langle T\rangle^f = \langle T\rangle_0^f = 300[\mathrm{K}] \tag{29}$$

$$\langle p\rangle^f = \langle p\rangle_0^f = 10^5[\mathrm{Pa}] \tag{30}$$

such that $\left\langle\rho_f\right\rangle^f = \left\langle\rho_f\right\rangle_0^f = \langle p\rangle_0^f/R\langle T\rangle_0^f = 10^5/(287\times 300) = 1.2[\mathrm{kg/m^3}]$

$$-\left(\varepsilon^*\left\langle k_f\right\rangle^f + \varepsilon k_{disxx}\right)\frac{d\langle T\rangle^f}{dx}\bigg|_{x=0} - \left((1-\varepsilon^*)k_s + \frac{16\sigma}{3\beta}\left(\langle T\rangle^s\right)^3\right)\frac{d\langle T\rangle^s}{dx}\bigg|_{x=0}$$
$$= I_0\cos\xi - (1-\varepsilon)\left(a\sigma\left(\left(\langle T\rangle_0^s\right)^4 - \left(\langle T\rangle_0^f\right)^4\right) + h_{conv}\left(\langle T\rangle_0^s - \langle T\rangle_0^f\right)\right) \tag{31}$$

where I_0 is the intensity of radiation and ξ is the incidence angle. Moreover, $a \cong 0.9$ is the emissivity of the front surface of the receiver, while h_{conv} is the convective heat transfer coefficient at the frontal surface. The properties of the air depend on the temperature, which makes the integrations of the foregoing governing equations formidable. In order to obtain analytic expressions for the unknown variables, we may approximate these properties by their representative values evaluated at the average air temperature over the receiver as given by

$$\overline{\langle T\rangle^f} = \frac{1}{L}\int_0^L \langle T\rangle^f dx \tag{32}$$

Likewise we shall define the solid phase average temperature as follows:

$$\overline{\langle T\rangle^s} = \frac{1}{L}\int_0^L \langle T\rangle^s dx \tag{33}$$

The two energy equations, that is, Eqs. (21) and (22) may be added together and integrated using the boundary conditions in Eqs. (29) and (31) to give

$$\begin{aligned} c_p G\left(\langle T\rangle^f - \langle T\rangle_0^f\right) &= \left(\varepsilon^* k_0 \left(\frac{\overline{\langle T\rangle^f}}{\langle T\rangle_0^f}\right)^n + \varepsilon k_{disxx}\right)\frac{d\langle T\rangle^f}{dx} \\ &+ \left((1-\varepsilon^*)k_s + \frac{16\sigma}{3\beta}\left(\overline{\langle T\rangle^s}\right)^3\right)\frac{d\langle T\rangle^s}{dx} + I_0\cos\xi \\ &- (1-\varepsilon)\left(a\sigma\left(\left(\langle T\rangle_0^s\right)^4 - \left(\langle T\rangle_0^f\right)^4\right) + h_{\mathrm{cov}}\left(\langle T\rangle_0^s - \langle T\rangle_0^f\right)\right) \end{aligned} \tag{34}$$

This equation is substituted into Eq. (21) to eliminate $\langle T\rangle^s$ in favor of $\langle T\rangle^f$. The resulting ordinary differential equation for $\langle T\rangle^f$ runs as

$$\begin{aligned} \frac{d^3\langle T\rangle^f}{dx^3} &= \frac{Gc_p}{\varepsilon^* k_0\left(\frac{\overline{\langle T\rangle^f}}{\langle T\rangle_0^f}\right)^n + \varepsilon k_{disxx}}\frac{d^2\langle T\rangle^f}{dx^2} \\ &+ h_v\frac{k_{stag} + \varepsilon k_{disxx} + \frac{16\sigma}{3\beta}\left(\overline{\langle T\rangle^s}\right)^3}{\left(\varepsilon^* k_0\left(\frac{\overline{\langle T\rangle^f}}{\langle T\rangle_0^f}\right)^n + \varepsilon k_{disxx}\right)\left((1-\varepsilon^*)k_s + \frac{16\sigma}{3\beta}\left(\overline{\langle T\rangle^s}\right)^3\right)}\frac{d\langle T\rangle^f}{dx} \\ &- h_v\frac{Gc_p}{\left(\varepsilon^* k_0\left(\frac{\overline{\langle T\rangle^f}}{\langle T\rangle_0^f}\right)^n + \varepsilon k_{disxx}\right)\left((1-\varepsilon^*)k_s + \frac{16\sigma}{3\beta}\left(\overline{\langle T\rangle^s}\right)^3\right)}\langle T\rangle^f \\ &+ h_v\frac{I_0\cos\xi - (1-\varepsilon)\left(a\sigma\left(\left(\langle T\rangle_0^s\right)^4 - \left(\langle T\rangle_0^f\right)^4\right) + h_{\mathrm{cov}}\left(\langle T\rangle_0^s - \langle T\rangle_0^f\right)\right)}{\left(\varepsilon^* k_0\left(\frac{\overline{\langle T\rangle^f}}{\langle T\rangle_0^f}\right)^n + \varepsilon k_{disxx}\right)\left((1-\varepsilon^*)k_s + \frac{16\sigma}{3\beta}\left(\overline{\langle T\rangle^s}\right)^3\right)} \end{aligned} \tag{35}$$

This ordinary differential equation, with the boundary conditions in Eqs. (29), (30) and (31) and the auxiliary asymptotic condition $d\langle T\rangle_0^f/dx = d^2\langle T\rangle_0^f/dx^2 = 0$, yields

$$\langle T\rangle^f = T_{eq} - \left(T_{eq} - \langle T\rangle_0^f\right)e^{-\gamma\lambda x} \tag{36}$$

and

$$\langle T\rangle^s = T_{eq} - \left(T_{eq} - \langle T\rangle_0^s\right)e^{-\gamma\lambda x} \tag{37}$$

where γ is the positive real root, which can uniquely be determined from the following cubic equation:

$$\gamma^3 + \frac{Gc_p}{\left(\varepsilon^* k_0\left(\frac{\overline{\langle T\rangle^f}}{\langle T\rangle_0^f}\right)^n + \varepsilon k_{disxx}\right)\lambda}\gamma^2 - \gamma - \frac{Gc_p}{\left(k_{stag} + \varepsilon k_{disxx} + \frac{16\sigma}{3\beta}\left(\overline{\langle T\rangle^s}\right)^3\right)\lambda} = 0 \tag{38}$$

where

$$\lambda = \sqrt{\frac{\left(k_{stag} + \varepsilon k_{disxx} + \frac{16\sigma}{3\beta}\left(\overline{\langle T\rangle^s}\right)^3\right)h_v}{\left(\varepsilon^* k_0\left(\frac{\overline{\langle T\rangle^f}}{\langle T\rangle_0^f}\right)^n + \varepsilon k_{disxx}\right)\left((1-\varepsilon^*)k_s + \frac{16\sigma}{3\beta}\left(\overline{\langle T\rangle^s}\right)^3\right)}} \tag{39}$$

The solid phase temperature at the inlet $\langle T\rangle_0^s$ and temperature at the thermal equilibrium, namely, $T_{eq} = \langle T\rangle_\infty^f = \langle T\rangle_\infty^s$, are given by

$$\langle T\rangle_0^s = T_{eq} + \left(T_{eq} - \langle T\rangle_0^f\right)\frac{Gc_p + \left(\varepsilon^* k_0\left(\frac{\overline{\langle T\rangle^f}}{\langle T\rangle_0^f}\right)^n + \varepsilon k_{disxx}\right)\gamma\lambda}{\left((1-\varepsilon^*)k_s + \frac{16\sigma}{3\beta}\left(\overline{\langle T\rangle^s}\right)^3\right)\gamma\lambda} \tag{40}$$

and

$$T_{eq} = \langle T\rangle_0^f + \frac{I_0\cos\xi - (1-\varepsilon)\left(a\sigma\left(\left(\langle T\rangle_0^s\right)^4 - \left(\langle T\rangle_0^f\right)^4\right) + h_{\text{cov}}\left(\langle T\rangle_0^s - \langle T\rangle_0^f\right)\right)}{Gc_p} \tag{41}$$

respectively. Usually, the receiver length L is sufficiently long to reach the local thermal equilibrium. Thus, the average air and solid temperatures are evaluated from

$$\overline{\langle T\rangle^f} = \frac{1-e^{-\gamma\lambda L}}{\gamma\lambda L}\langle T\rangle_0^f + \left(1 - \frac{1-e^{-\gamma\lambda L}}{\gamma\lambda L}\right)T_{eq} \cong \frac{1}{\gamma\lambda L}\langle T\rangle_0^f + \left(1 - \frac{1}{\gamma\lambda L}\right)T_{eq} \tag{42}$$

$$\overline{\langle T\rangle^s} = \frac{1-e^{-\gamma\lambda L}}{\gamma\lambda L}\langle T\rangle_0^s + \left(1 - \frac{1-e^{-\gamma\lambda L}}{\gamma\lambda L}\right)T_{eq} \cong \frac{1}{\gamma\lambda L}\langle T\rangle_0^s + \left(1 - \frac{1}{\gamma\lambda L}\right)T_{eq} \tag{43}$$

As one of the most important performance parameters, the receiver efficiency is defined by

$$\eta = \frac{I_0 \cos\xi - (1-\varepsilon)\left(a\sigma\left(\left(\langle T\rangle_0^s\right)^4 - \left(\langle T\rangle_0^f\right)^4\right) + h_{\text{cov}}\left(\langle T\rangle_0^s - \langle T\rangle_0^f\right)\right)}{I_0 \cos\xi} \tag{44}$$

Having established the temperature development, the momentum equation, that is, Eq. (20) along with the equation of state can easily be solved to find out the pressure distribution along the receiver as

$$\langle p\rangle^f = \sqrt{\left(\langle p\rangle_0^f\right)^2 - 2R\left(\frac{\mu_0}{K}\left(\frac{\langle T\rangle^f}{\langle T\rangle_0^f}\right)^n G + bG^2\right)\left(\frac{1-e^{-\gamma\lambda x}}{\gamma\lambda}\langle T\rangle_0^f + \left(x - \frac{1-e^{-\gamma\lambda x}}{\gamma\lambda}\right)T_{eq}\right)} \tag{45}$$

Under the low Mach approximation, the required pumping power per unit frontal area may be evaluated from

$$PP = G\int_0^L -\frac{d\langle p\rangle^f}{\left\langle\rho_f\right\rangle^f} = G\int_0^L -\frac{1}{\left\langle\rho_f\right\rangle^f}\frac{d\langle p\rangle^f}{dx}dx$$

$$= \left(\frac{R}{\langle p\rangle_0^f}\right)^2 G\left(\frac{\mu_0}{K}\left(\frac{\langle T\rangle^f}{\langle T\rangle_0^f}\right)^n G + bG^2\right)\int_0^L \left(\langle T\rangle^f\right)^2 dx$$

$$= \left(\frac{R}{\langle p\rangle_0^f}\right)^2 G\left(\frac{\mu_0}{K}\left(\frac{\langle T\rangle^f}{\langle T\rangle_0^f}\right)^n G + bG^2\right)L\left(T_{eq}{}^2 - \frac{2\left(1-e^{-\gamma\lambda L}\right)}{\gamma\lambda L}T_{eq}\left(T_{eq} - \langle T\rangle_0^f\right)\right.$$

$$\left. + \frac{1-e^{-2\gamma\lambda L}}{2\gamma\lambda L}\left(T_{eq} - \langle T\rangle_0^f\right)^2\right)$$

$$\cong \frac{G}{\left(\langle\rho\rangle_0^f\right)^2}\left(\frac{\mu_0}{K}\left(\frac{\langle T\rangle^f}{\langle T\rangle_0^f}\right)^n G + bG^2\right)\frac{1}{2\gamma\lambda}\left((2\gamma\lambda L - 3)\left(\frac{T_{eq}}{\langle T\rangle_0^f}\right)^2 + 2\left(\frac{T_{eq}}{\langle T\rangle_0^f}\right) + 1\right) \tag{46}$$

Note that the dynamic pressure change is sufficiently small as compared to the absolute pressure such that $\left\langle\rho_f\right\rangle^f \propto 1/\langle T\rangle^f$.

3.2 Analysis based on the P1 model

Since the Rosseland approximation used in the previous analysis ceases to be valid near boundaries, the validity of applying the Rosseland approximation near the inlet boundary of the receiver should be investigated. In order to examine the validity of the Rosseland approximation, the results based on the Rosseland approximation will be compared with the results obtained from solving the irradiation transport equation based on the P1 model. Since the silicon carbide ceramic foam is optically thick, the radiant energy emitted from other locations in the domain is quickly absorbed such that the radiative heat flux is given by

$$q_R = -\frac{1}{3\beta}\frac{\partial G}{\partial x_j} \tag{47}$$

where the diffuse integrated intensity G_r satisfies the irradiation transport equation based on the P1 model as follows:

$$\frac{\partial}{\partial x_j}\left(\frac{1}{3\beta}\frac{\partial G_r}{\partial x_j}\right) + \kappa\left(4\sigma\left(\langle T\rangle^s\right)^4 - G_r\right) = 0 \tag{48}$$

where κ is the absorption coefficient.

Moreover, the effects of turbulence mixing on the heat transfer are also considered. Therefore, the energy equation for the air will be written as

$$\varepsilon c_p \frac{\partial}{\partial x_j}\left\langle \rho_f \right\rangle^f \langle u_j\rangle^f \langle T\rangle^f = \frac{\partial}{\partial x_j}\left(\varepsilon^* \left\langle k_f\right\rangle^f \frac{\partial\langle T\rangle^f}{\partial x_j} + \varepsilon\left(k_{dis_{jk}} + \frac{c_{p_f}\langle\mu_t\rangle^f}{\sigma_T}\right)\frac{\partial\langle T\rangle^f}{\partial x_k}\delta_{jk}\right) - h_v\left(\langle T\rangle^f - \langle T\rangle^s\right) \tag{49}$$

where turbulent Prandtl number $\sigma_T = 0.9$ is assumed to be constant.

Under the low Mach number approximation, namely, we may reduce the macroscopic governing equations namely Eqs. (1), (49), (14) and (48) to a one-dimensional set of the equations as follows:

$$-\frac{d\langle p\rangle^f}{dx} = \frac{\langle\mu\rangle^f}{K}\frac{G}{\left\langle \rho_f\right\rangle^f} + b\frac{G^2}{\left\langle \rho_f\right\rangle^f} = \frac{R}{\langle p\rangle^f}\left(\frac{\langle\mu\rangle^f}{K}G + bG^2\right)\langle T\rangle^f \tag{50}$$

$$c_p G\frac{d\langle T\rangle^f}{dx} = \frac{d}{dx}\left(\varepsilon^* \left\langle k_f\right\rangle^f + \varepsilon\left(k_{dis_{xx}} + \frac{c_{p_f}\langle\mu_t\rangle^f}{\sigma_T}\right)\right)\frac{d\langle T\rangle^f}{dx} - h_v\left(\langle T\rangle^f - \langle T\rangle^s\right) \tag{51}$$

$$\frac{d}{dx}\left((1-\varepsilon^*)k_s\frac{d\langle T\rangle^s}{dx} + \frac{1}{3\beta}\frac{dG_r}{dx}\right) - h_v\left(\langle T\rangle^s - \langle T\rangle^f\right) = 0 \tag{52}$$

$$\frac{d}{dx}\left(\frac{1}{3\beta}\frac{dG_r}{dx}\right) + \kappa\left(4\sigma\left(\langle T\rangle^s\right)^4 - G_r\right) = 0 \tag{53}$$

The turbulence kinetic energy is dropped from the momentum equation since it stays nearly constant within the receiver.

Nakayama and Kuwahara [22] established the macroscopic two-equation turbulence model, which does not require any detailed morphological information for the structure. The model, for given permeability and Forchheimer coefficient, can be used for analyzing most complex turbulent flow situations in homogeneous porous media. For the case of fully developed turbulent flow in an isotropic porous structure, the eddy viscosity is given by

$$\langle \mu_t \rangle^f = 2GbK \tag{54}$$

Note that

$$k_{dis_{xx}} / \left(c_{p_f} \langle \mu_t \rangle^f / \sigma_T \right) = 0.6\sigma_T / \varepsilon b \sqrt{K} > 1$$

such that the dispersion thermal conductivity usually overwhelms the eddy thermal conductivity.

For absorption coefficient κ, the measurements made on cordierite ceramic foams by Kamiuto et al. [22] give the following correlation:

$$\kappa = 4a(1-\varepsilon)/d_m \tag{55}$$

The boundary conditions of $\langle T \rangle^f$ and $\langle p \rangle^f$ are the same as Eqs. (29) and (30). The other boundary conditions are given as follows:

$$q_{R_x} = -\frac{1}{3\beta}\frac{dG_r}{dx} = -\frac{G_r}{2} \tag{56}$$

and

$$\begin{aligned} &-\left(\varepsilon^* \langle k_f \rangle^f + \varepsilon \left(k_{disxx} + \frac{c_{p_f} \langle \mu_t \rangle^f}{\sigma_T} \right) \right) \frac{d\langle T \rangle^f}{dx}\bigg|_{x=0} - (1-\varepsilon^*) k_s \frac{d\langle T \rangle^s}{dx}\bigg|_{x=0} - \frac{G_r|_{x=0}}{2} \\ &= (1-(1-a)(1-\varepsilon)) I_0 \cos\xi \\ &\quad - (1-\varepsilon)\left(a\sigma\left((\langle T \rangle_0^s)^4 - \left(\langle T \rangle_0^f\right)^4 \right) + h_{conv}\left(\langle T \rangle_0^s - \langle T \rangle_0^f \right) \right) \end{aligned} \tag{57}$$

Furthermore, the streamwise gradients of the dependent variables $\langle T \rangle^f$, $\langle T \rangle^s$ and G_r are set to zero sufficiently far downstream at $x = L$.

The two energy equations, namely, Eqs. (51) and (52) may be added together and integrated using the boundary conditions in Eqs. (29) and (57) to give

$$\begin{aligned} &c_p G\left(\langle T \rangle^f - \langle T \rangle_0^f \right) \\ &= \left(\varepsilon^* k_0 \left(\frac{\langle T \rangle^f}{\langle T \rangle_0^f} \right)^n + \varepsilon \left(k_{dis_{xx}} + \frac{c_{p_f} \langle \mu_t \rangle^f}{\sigma_T} \right) \right) \frac{d\langle T \rangle^f}{dx} + (1-\varepsilon^*) k_s \frac{\partial \langle T \rangle^s}{\partial x_j} + \frac{1}{3\beta}\frac{dG_r}{dx} \\ &+ (1-(1-a)(1-\varepsilon)) I_0 \cos\xi - (1-\varepsilon)\left(a\sigma\left((\langle T \rangle_0^s)^4 - \left(\langle T \rangle_0^f\right)^4 \right) + h_{conv}\left(\langle T \rangle_0^s - \langle T \rangle_0^f \right) \right) \end{aligned} \tag{58}$$

Eqs. (52) and (53) are combined to give

$$G_r = 4\sigma(\langle T \rangle^s)^4 + \frac{h_v}{\kappa}\left(\langle T \rangle^s - \langle T \rangle^f \right) - \frac{(1-\varepsilon^*) k_s}{\kappa} \frac{d^2 \langle T \rangle^s}{dx^2} \tag{59}$$

This equation, Eq. (59), and Eq. (51) are substituted into Eq. (58) to eliminate G_r and $\langle T \rangle^s$ in favor of $\langle T \rangle^f$. The resulting ordinary differential equation for $\langle T \rangle^f$ runs as

$$
\begin{aligned}
\frac{d^5\langle T\rangle^f}{dx^5} &= \frac{Gc_p}{\varepsilon^* k_0\left(\frac{\overline{\langle T\rangle^f}}{\langle T\rangle_0^f}\right)^n+\varepsilon\left(k_{dis_{xx}}+\frac{c_{p_f}\langle\mu_t\rangle^f}{\sigma_T}\right)}\frac{d^4\langle T\rangle^f}{dx^4} \\
&+\left(3\beta\kappa\frac{(1-\varepsilon^*)k_s+\frac{16\sigma}{3\beta}\left(\overline{\langle T\rangle^s}\right)^3+\frac{h_v}{3\beta\kappa}}{(1-\varepsilon^*)k_s}+\frac{h_v}{\varepsilon^* k_0\left(\frac{\overline{\langle T\rangle^f}}{\langle T\rangle_0^f}\right)^n+\varepsilon\left(k_{dis_{xx}}+\frac{c_{p_f}\langle\mu_t\rangle^f}{\sigma_T}\right)}\right)\frac{d^3\langle T\rangle^f}{dx^3} \\
&-\left((1-\varepsilon^*)k_s+\frac{16\sigma}{3\beta}\left(\overline{\langle T\rangle^s}\right)^3+\frac{h_v}{3\beta\kappa}\right)\frac{3\beta\kappa Gc_p}{\left(\varepsilon^* k_0\left(\frac{\overline{\langle T\rangle^f}}{\langle T\rangle_0^f}\right)^n+\varepsilon\left(k_{dis_{xx}}+\frac{c_{p_f}\langle\mu_t\rangle^f}{\sigma_T}\right)\right)(1-\varepsilon^*)k_s}\frac{d^2\langle T\rangle^f}{dx^2} \\
&-3\beta\kappa h_v\frac{k_{stag}+\varepsilon\left(k_{dis_{jk}}+\frac{c_{p_f}\langle\mu_t\rangle^f}{\sigma_T}\right)+\frac{16\sigma}{3\beta}\left(\overline{\langle T\rangle^s}\right)^3}{\left(\varepsilon^* k_0\left(\frac{\overline{\langle T\rangle^f}}{\langle T\rangle_0^f}\right)^n+\varepsilon\left(k_{dis_{xx}}+\frac{c_{p_f}\langle\mu_t\rangle^f}{\sigma_T}\right)\right)(1-\varepsilon^*)k_s}\frac{d\langle T\rangle^f}{dx} \\
&+\frac{3\beta\kappa h_v Gc_p}{\left(\varepsilon^* k_0\left(\frac{\overline{\langle T\rangle^f}}{\langle T\rangle_0^f}\right)^n+\varepsilon\left(k_{dis_{xx}}+\frac{c_{p_f}\langle\mu_t\rangle^f}{\sigma_T}\right)\right)(1-\varepsilon^*)k_s}\left(\langle T\rangle^f-\langle T\rangle_0^f\right) \\
&-3\beta\kappa h_v\frac{(1-(1-a)(1-\varepsilon))I_0\cos\xi-(1-\varepsilon)\left(a\sigma\left(\left(\langle T\rangle_0^s\right)^4-\left(\langle T\rangle_0^f\right)^4\right)+h_{conv}\left(\langle T\rangle_0^s-\langle T\rangle_0^f\right)\right)}{\left(\varepsilon^* k_0\left(\frac{\overline{\langle T\rangle^f}}{\langle T\rangle_0^f}\right)^n+\varepsilon\left(k_{dis_{xx}}+\frac{c_{p_f}\langle\mu_t\rangle^f}{\sigma_T}\right)\right)(1-\varepsilon^*)k_s}
\end{aligned}
\tag{60}
$$

This ordinary differential equation, with the boundary conditions in Eqs. (29), (56) and (57) and the zero derivative conditions far downstream ($x \to \infty$: Note L is sufficiently large), yields Eqs. (36) and (37). Note that γ is the positive real root which can be determined from the following characteristic equation:

$$
\begin{aligned}
&\gamma^5+\frac{Gc_p}{\left(\varepsilon^* k_0\left(\frac{\overline{\langle T\rangle^f}}{\langle T\rangle_0^f}\right)^n+\varepsilon\left(k_{dis_{xx}}+\frac{c_{p_f}\langle\mu_t\rangle^f}{\sigma_T}\right)\right)\lambda}\gamma^4 \\
&-\frac{\frac{3\beta\kappa}{h_v}\left(\varepsilon^* k_0\left(\frac{\overline{\langle T\rangle^f}}{\langle T\rangle_0^f}\right)^n+\varepsilon\left(k_{dis_{xx}}+\frac{c_{p_f}\langle\mu_t\rangle^f}{\sigma_T}\right)\right)\left((1-\varepsilon^*)k_s+\frac{16\sigma}{3\beta}\left(\overline{\langle T\rangle^s}\right)^3+\frac{h_v}{3\beta\kappa}\right)+(1-\varepsilon^*)k_s}{k_{stag}+\varepsilon\left(k_{dis_{xx}}+\frac{c_{p_f}\langle\mu_t\rangle^f}{\sigma_T}\right)+\frac{16\sigma}{3\beta}\left(\overline{\langle T\rangle^s}\right)^3}\gamma^3 \\
&-\frac{(1-\varepsilon^*)k_s+\frac{16\sigma}{3\beta}\left(\overline{\langle T\rangle^s}\right)^3+\frac{h_v}{3\beta\kappa}}{k_{stag}+\varepsilon\left(k_{dis_{xx}}+\frac{c_{p_f}\langle\mu_t\rangle^f}{\sigma_T}\right)+\frac{16\sigma}{3\beta}\left(\overline{\langle T\rangle^s}\right)^3}\frac{3\beta\kappa Gc_p}{h_v\lambda}\gamma^2+\frac{3\beta\kappa}{\lambda^2}\gamma \\
&+\frac{3\beta\kappa Gc_p}{\left(k_{stag}+\varepsilon\left(k_{dis_{xx}}+\frac{c_{p_f}\langle\mu_t\rangle^f}{\sigma_T}\right)+\frac{16\sigma}{3\beta}\left(\overline{\langle T\rangle^s}\right)^3\right)\lambda^3}=0
\end{aligned}
\tag{61}
$$

where

$$\lambda = \sqrt{\frac{\left(k_{stag} + \varepsilon\left(k_{dis_{xx}} + \frac{c_{p_f}\langle\mu_t\rangle^f}{\sigma_T}\right) + \frac{16\sigma}{3\beta}\left(\overline{\langle T\rangle^s}\right)^3\right)h_v}{\left(\varepsilon^* k_0\left(\frac{\overline{\langle T\rangle^f}}{\langle T\rangle_0^f}\right)^n + \varepsilon\left(k_{dis_{xx}} + \frac{c_{p_f}\langle\mu_t\rangle^f}{\sigma_T}\right)\right)(1-\varepsilon^*)k_s}} \tag{62}$$

The solid phase temperature at the inlet $\langle T\rangle_0^s$ and temperature at the thermal equilibrium, namely,$T_{eq} = \langle T\rangle_\infty^f = \langle T\rangle_\infty^s$, are determined from the following implicit equations:

$$\begin{aligned} Gc_p\left(T_{eq} - \langle T\rangle_0^f\right) = &-\left(\varepsilon^* k_0 + \varepsilon\left(k_{dis_{xx}} + \frac{c_{p_f}\langle\mu_t\rangle^f}{\sigma_T}\right)\right)\gamma\lambda\left(T_{eq} - \langle T\rangle_0^f\right) \\ &-(1-\varepsilon^*)k_s\gamma\lambda\left(T_{eq} - \langle T\rangle_0^s\right) \\ &-\frac{1}{2}\left(4\sigma\left(\langle T\rangle_0^s\right)^4 + \frac{h_v}{\kappa}\left(\langle T\rangle_0^s - \langle T\rangle_0^f\right) + \frac{(1-\varepsilon^*)k_s}{\kappa}(\gamma\lambda)^2\left(T_{eq} - \langle T\rangle_0^s\right)\right) \end{aligned} \tag{63}$$

$$\begin{aligned} T_{eq} = &\langle T\rangle_0^f \\ &+ \frac{(1-(1-a)(1-\varepsilon))I_0\cos\xi - (1-\varepsilon)\left(a\sigma\left(\left(\langle T\rangle_0^s\right)^4 - \left(\langle T\rangle_0^f\right)^4\right) + h_{conv}\left(\langle T\rangle_0^s - \langle T\rangle_0^f\right)\right)}{Gc_p} \end{aligned} \tag{64}$$

where the boundary condition in Eq. (56) is utilized. Usually, the receiver length L is sufficiently long to reach the local thermal equilibrium. Thus, the average air and solid temperatures are evaluated from Eqs. (42) and (43).

4. Validations of the Rosseland approximation

Smirnova et al. [23] numerically studied the compressible fluid flow and heat transfer within the solar receiver with silicon carbide monolithic honeycombs. In their paper, the following input data were collected to obtain the analytic solutions based on the present local thermal non-equilibrium model:

$$\begin{aligned} &\left\langle\rho_f\right\rangle_0^f = 1.2[\mathrm{kg/m^3}],\ \langle T\rangle_0^f = 300[\mathrm{K}](\langle p\rangle_0^f = 10^5[\mathrm{Pa}]),\ c_p = 1000[\mathrm{J/kgK}], \\ &G = 1.2[\mathrm{kg/m^2s}],\ L = 0.05[m],\ I_0 = 10^6[\mathrm{W/m^2}],\ \xi = 0,\ h_{conv} = 0[\mathrm{W/m^2K}], \\ &k_s = 150[\mathrm{W/mK}]k_{dis} = 0[\mathrm{W/mK}],\ h_v = 8.8\times10^4[\mathrm{W/m^3K}],\ \varepsilon = 0.5,\ \beta = 50[1/\mathrm{m}]. \end{aligned}$$

However, it should be noticed that the porosity of the silicon carbide monolithic honeycombs is not available in Smirnova et al. [23], its value was estimated to be $\varepsilon = 0.5$ from the figure provided by Agrafiotis et al. [24]. The mean extinction coefficient β for silicon carbide monolithic honeycombs is not available in their paper. Finally, the value was estimated to be 50[1/m] by correlating the present results against theirs. It should also be noted that the convective heat transfer coefficient was set to zero since radiation predominates over convection in the receiver front.

As for possible instabilities, the fifth-order characteristic Eq. (61) based on the P1 model should be examined carefully. **Figure 2** shows the residual of the fifth-order characteristic equation $f_R(\gamma)$. The figure clearly shows that the fifth-order

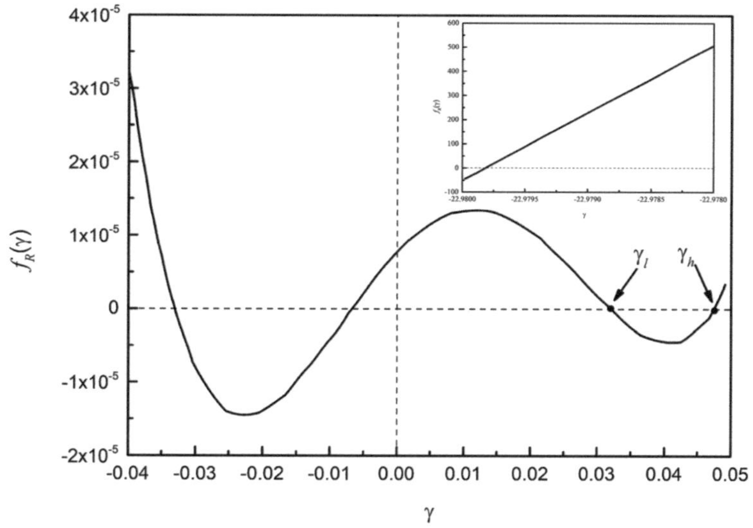

Figure 2.
Residual of the fifth-order characteristic equation.

characteristic Eq. (61) under a possible range of the silicon carbide parameters yields two positive roots γ_h and γ_l, which are fairly close to each other. The corresponding temperature variations of both phases however depend strongly on its value, which results in a non-unique value of equilibrium temperature. Since flow instability is inferred by an unexpected nature of the quadratic pressure difference with respect to equilibrium temperature, the existence of two positive roots may be responsible for possible hydrodynamic and thermal instabilities reported previously. A further investigation based on an unsteady procedure is definitely needed to explore possible causes of these instabilities, closely related to the radiative heat transfer mode.

The third-order characteristic Eq. (38) based on the Rosseland approximation, on the other hand, yields only one positive root γ_1. The corresponding fluid and

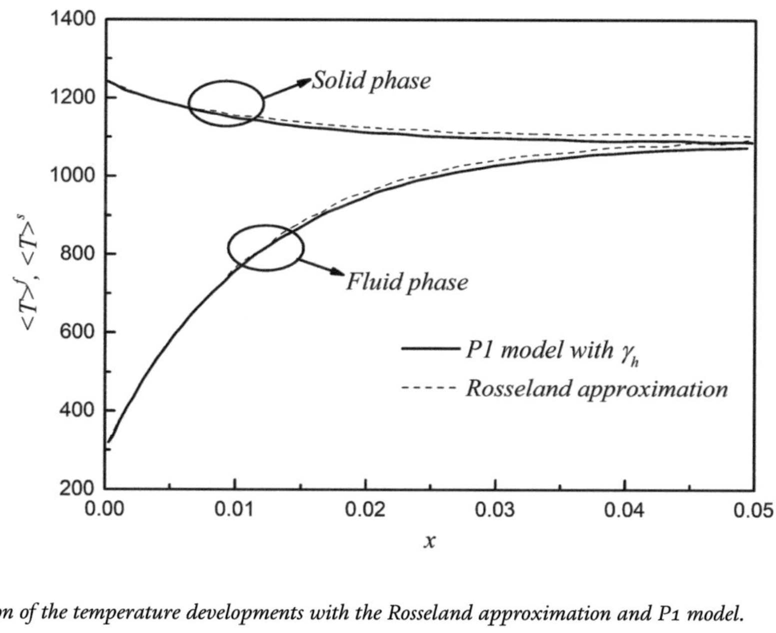

Figure 3.
Comparison of the temperature developments with the Rosseland approximation and P1 model.

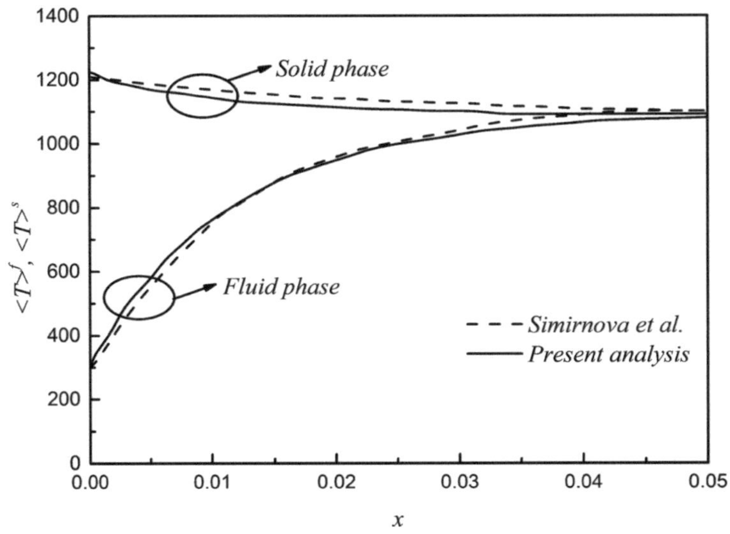

Figure 4.
Axial developments of the fluid and solid phase temperatures: comparison of the present analysis and FEM analysis.

solid temperature variations generated under the Rosseland approximation are compared with those based on the P1 model with the larger root γ_h. **Figure 3** shows that both sets of the temperature developments agree fairly well with each other. Thus, the Rosseland approximation for this case, despite its failure near the inlet boundary, is fairly accurate and may well be used for quick estimations and further analysis.

In **Figure 4**, the present analytic solutions are compared against the large-scale FEM numerical calculations based on COMSOL, reported by Smirnova et al. [23]. It should be mentioned that the direct numerical integrations of Eqs. (20)–(22) were also carried out using the finite volume method code, SAINTS [12]. As the convergence criteria, the residuals of all equations are less than 10^{-5}. It can be clearly seen that the air temperature increases as receiving heat from the monolithic receiver. Eventually, these two phases reach local thermal equilibrium near the exit. Both sets of solutions agree very well with each other, indicating the validity of the present local thermal non-equilibrium model.

5. Applications to silicon carbide ceramic foam volumetric receiver

In order to overcome the problems associated with thermal spots and flow instabilities, we would like to study fluid flow and heat transfer characteristics in silicon carbide ceramic foams based on the analytical expressions of pressure and temperature fields within a solar volumetric receiver. The performance of the receiver may be assessed in terms of the receiver efficiency η under equal pumping powerPP. Thus, the effects of the pore diameter d_m on the receiver efficiency η are presented in **Figure 5**, since d_m is a crucial geometry parameter affecting hydrodynamic and thermal characteristics of foam shown in Eqs. (23), (24) and (27). The pore diameter d_m is varied whereas the other parameters are fixed as follows:

$$\left\langle \rho_f \right\rangle_0^f = 1.2[\mathrm{kg/m^3}],\ \langle T\rangle_0^f = 300[\mathrm{K}](\langle p\rangle_0^f = 10^5[\mathrm{Pa}]),\ c_p = 1000[\mathrm{J/kgK}],$$
$$L = 0.03[m],\ I_0 = 10^6[\mathrm{W/m^2}],\ \xi = 0, h_{conv} = 0[\mathrm{W/m^2K}],\ k_s = 150[\mathrm{W/mK}],\ \varepsilon = 0.9.$$

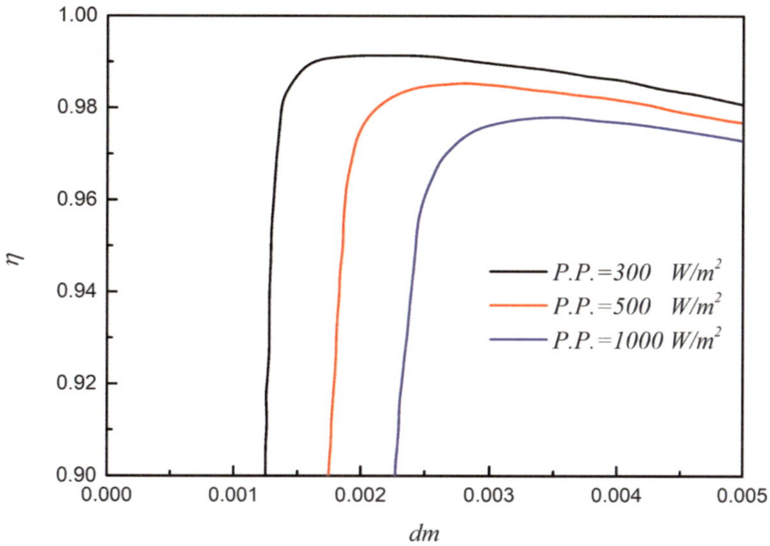

Figure 5.
Effects of the pore diameter on the receiver efficiency.

All other parameters are evaluated using Eqs. (17), (18) and from Eq. (23) to (28).

As shown in **Figure 5**, it is interesting to note that η suddenly increases at some critical value of d_m for a given value of PP, which means that the pore diameter d_m must be larger than this critical value to achieve high η. This finding is useful to design a volumetric receiver, and can be interpreted in what follows.

As indicated in Eq. (46), it can be easily deduced that $G \propto \sqrt{PP}$ for low PP and $G \propto \sqrt[3]{PP}$ for high PP, which results in that the amount of heat carried by the air, $G\left(T_{eq} - \langle T \rangle_0^f\right) \propto \sqrt{PP}$, increases drastically on increasing the pumping power PP from zero. Nevertheless, its rate of increase diminishes for the higher PP range, in which $G\left(T_{eq} - \langle T \rangle_0^f\right) \propto \sqrt[3]{PP}$. Moreover, it can also be concluded that the sharp rise in the receiver efficiency occurs around the transition from the Darcy to Forchheimer regime, namely,

$$\frac{\mu_0}{K}\left(\frac{T_{eq}}{\langle T \rangle_0^f}\right)^n G_{tr} \cong b{G_{tr}}^2 \tag{65}$$

or

$$G_{tr} \cong \left(\frac{\mu_0}{bK}\left(\frac{I_0 \cos \xi}{c_p \langle T \rangle_0^f}\right)^n\right)^{\frac{1}{1+n}} \tag{66}$$

since

$$\frac{T_{eq}}{\langle T \rangle_0^f} \cong \frac{c_p G_{tr} \langle T \rangle_0^f + I_0 \cos \xi}{c_p G_{tr} \langle T \rangle_0^f} \cong \frac{I_0 \cos \xi}{c_p G_{tr} \langle T \rangle_0^f} \tag{67}$$

Thus, Eq. (46) may be written for the case in which the sharp rise in η takes place as follows:

$$PP = \frac{G_{tr}}{\left(\langle\rho\rangle_0^f\right)^2}\left(2bG_{tr}{}^2\right)\left(\frac{T_{eq}}{\langle T\rangle_0^f}\right)^2 L \cong \frac{2bLG_{tr}{}^3}{\left(\langle\rho\rangle_0^f\right)^2}\left(\frac{I_0\cos\xi}{c_pG_{tr}\langle T\rangle_0^f}\right)^2$$
$$= \frac{2bL}{\left(\langle\rho\rangle_0^f\right)^2}\left(\frac{I_0\cos\xi}{c_p\langle T\rangle_0^f}\right)^2\left(\frac{\mu_0}{bK}\left(\frac{I_0\cos\xi}{c_p\langle T\rangle_0^f}\right)^n\right)^{\frac{1}{1+n}} \tag{68}$$

which, for given PP, gives the minimum value of the pore diameter d_{mtr}:

$$\frac{d_{mtr}}{L} = f(\varepsilon)\left(\frac{2}{\left(\langle\rho\rangle_0^f\right)^2 PP}\left(\frac{I_0\cos\xi}{c_p\langle T\rangle_0^f}\right)^3\right)^{\frac{1+n}{2+n}}\left(\frac{\mu_0}{\left(\frac{I_0\cos\xi}{c_p\langle T\rangle_0^f}\right)L}\right)^{\frac{1}{2+n}} \tag{69}$$

$$f(\varepsilon) = \left(\frac{(bd_m)^n}{K/d_m{}^2}\right)^{\frac{1}{2+n}} = \left(\frac{(12(1-\varepsilon))^n}{0.00073(1-\varepsilon)^{-0.224}\left(\frac{1.18}{1-e^{-(1-\varepsilon)/0.04}}\sqrt{\frac{1-\varepsilon}{3\pi}}\right)^{-1.11}}\right)^{\frac{1}{2+n}} \tag{70}$$

For PP = 300, 500 and 1000 W/m^2 studied here, Eq. (69) gives d_{mtr}= 0.0022, 0.0016 and 0.0010 m, respectively. It is consistent with what is observed in **Figure 5**, since an increase in d_m (i.e., decrease inβ) from d_{mtr} makes further penetration of the solar radiation possible. This works to keep the solid temperature at the inlet comparatively low such that heat loss to the ambient by radiation is suppressed. As a result, high receiver efficiency can be achieved. However, the increase in d_m on the other hand results in decreasing the volumetric heat transfer coefficient, as can be seen from Eq. (27). Too large d_m deteriorates interstitial heat transfer from the solid to air. Thus, as can be seen from the figure, the optimal size of d_m exits under the equal pumping power constraint.

In order to achieve local thermal equilibrium for the two phases within the receiver, the length of the receiver is assumed to be sufficiently long in the present study. In view of minimizing the required pumping power, however, it is noticeable that shorter length is better, as clearly seen from Eq. (46). Hence, a minimum length required to approach local thermal equilibrium may be chosen to design a receiver, which would guarantee both maximum receiver efficiency and minimum pumping power. Therefore, we may roughly set the optimal receiver length as

$$L = \frac{3}{\gamma\lambda} \tag{71}$$

such that

$$\frac{\langle T\rangle^f\Big|_{x=L} - T_{eq}}{\langle T\rangle_0^f - T_{eq}} = \frac{\langle T\rangle^s\big|_{x=L} - T_{eq}}{\langle T\rangle_0^s - T_{eq}} = e^{-3} \cong 5\% \tag{72}$$

Eq. (71) together with Eq. (69) provides useful information for designing a volumetric solar receiver of silicon carbide ceramic foam.

6. Conclusions

For the first time, the complete set of analytical solutions, which fully considers the combined effects of turbulence, tortuosity, thermal dispersion, compressibility on the convective, conductive and radiative heat transfer within a ceramic foam receiver, is presented based on the two-energy equation model of porous media. Both the Rosseland approximation and the P1 model are applied to account for the radiative heat transfer through the solar receiver, while the low Mach approximation is exploited to investigate the compressible flow through the receiver. Based on the P1 model, two positive roots were found from the characteristic equations of the fifth-order differential equation, indicating possible occurrence of hydrodynamic and thermal instabilities. However, it has been found that the Rosseland approximation for this case, despite its failure near the inlet boundary, is fairly accurate and may well be used for quick estimations and further analysis. Due to their advantages, such as high thermal conductivity and fluid mixing, silicon carbide ceramic foams are considered as a possible candidate for the receiver, which can overcome the problems associated with thermal spots and flow instabilities. The results show that the pore diameter must be larger than its critical value to achieve high receiver efficiency. As a result, there exists an optimal pore diameter for achieving the maximum receiver efficiency under the equal pumping power. The optimal pore diameter yielding the maximum receiver efficiency may be found around the critical value given by Eq. (71). A simple relation is derived for determining the length of the volumetric solar receivers of silicon carbide ceramic foam.

Conflict of interest

The authors declare no conflict of interest.

Nomenclature

A	surface area (m^2)
A_{int}	interfacial surface area between the fluid and solid (m^2)
b	inertial coefficient (1/m)
c	specific heat (J/kg K)
cp	specific heat at constant pressure (J/kg K)
d_m	pore diameter (m)
G	mass flux (kg/m^2 s)
h	specific enthalpy (J/kg)
h_v	volumetric heat transfer coefficient (W/m^3K)
I_0	intensity of radiation (W/m^2)
k	thermal conductivity (W/m K)
K	permeability (m^2)
L	receiver length (m)
n_j	normal unit vector from the fluid side to solid side (−)
PP	pumping power per unit frontal area (W/m^2)
Pr	Prandtl number (−)
q	heat flux (W/m^2)
R	gas constant (J/kg K)
T	temperature (K)
u_i	velocity vector (m/s)
V	representative elementary volume (m^3)
x_i	Cartesian coordinates (m)
x	axial coordinate (m)

β	mean extinction coefficient (1/m)
γ	dimensionless parameter (−)
ε	porosity (−)
ε^*	effective porosity (−)
ξ	incidence angle (rad)
η	receiver efficiency (−)
λ	characteristic coefficient (1/m)
μ	viscosity (Pa s)
ν	kinematic viscosity (m^2/s)
ρ	density (kg/m^2)
σ	Stephan-Boltzmann constant (W/m^2K^4)
κ	absorption coefficient (1/m)
τ_{ij}	stress tensor (Pa)

Special symbols

$\tilde{\varphi}$	deviation from intrinsic average
$\langle\phi\rangle$	Darcian average
$\langle\phi\rangle^{f,s}$	intrinsic average

Subscripts and superscripts

dis	dispersion
eq	equilibrium
f	fluid
s	solid
stag	stagnation
0	reference

Author details

Chen Yang[1], Huijin Xu[2]* and Akira Nakayama[3]

1 College of Chemical Engineering, Fuzhou University, China

2 China-UK Low Carbon College, Shanghai Jiao Tong University, China

3 Faculty of Engineering, Shizuoka University, Japan

*Address all correspondence to: xuhuijin@sjtu.edu.cn

References

[1] Fend T. High porosity materials as volumetric receivers for solar energetics. Optica Applicata. 2010; **40**(2):271-284

[2] Alexoprrroulos S, Hoffschmidt B. Solar tower power plant in Germany and future perspectives of the development of the technology in Greece and Cyprus. Renewable Energy. 2010;**35**:1352-1356

[3] Pitz-Paal R, Hoffschmidt B, Bohmer M, Becker M. Experimental and numerical evaluation of the performance and flow stability of different types of open volumetric absorbers under non-homogeneous irradiation. Solar Energy. 1997;**60**:135-150

[4] Becker M, Fend T, Hoffschmidt B, Pitz-Paal R, Reutter O, Stamatov V, et al. Theoretical and numerical investigation of flow stability in porous materials applied as volumetric solar receiver. Solar Energy. 2006;**80**: 1241-1248

[5] Fend T, Hoffschmidt B, Pitz-Paal R, Reutter O. Porous materials as open volumetric solar receivers: Experimental determination of thermophysical and heat transfer properties. Energy. 2004; **29**:823-833

[6] Bai F. One dimensional thermal analysis of silicon carbide ceramic foam used for solar air receiver. International Journal of Thermal Sciences. 2010;**49**: 2400-2404

[7] Sano Y, Iwase S, Nakayama A. A local thermal non-equilibrium analysis of silicon carbide ceramic foam as a solar volumetric receiver. Journal of Solar Energy Engineering, Transactions of the ASME. 2012;**134**(2):021006

[8] Wu Z, Caliot C, Flamant G, Wang Z. Numerical simulation of convective heat transfer between air flow and ceramic foams to optimize volumetric solar air receiver performances. International Journal of Heat and Mass Transfer. 2011; **54**:1527-1537

[9] Yang C, Nakayama A. A synthesis of tortuosity and dispersion in effective thermal conductivity of porous media. International Journal of Heat and Mass Transfer. 2010;**53**(15–16):3222-3230

[10] Neuman SP. Theoretical derivation of Darcy's law. Acta Mechanica. 1977;**25**: 153-170

[11] Cheng P. Heat transfer in geothermal systems. Advances in Heat Transfer. 1978;**14**:1-105

[12] Nakayama A. PC-Aided Numerical Heat Transfer and Convective Flow. CRC Press; 1995. pp. 49-50, 103-115

[13] Quintard M, Whitaker S. One and two equation models for transient diffusion processes in two-phase systems. Advances in Heat Transfer. 1993;**23**:369-465

[14] Nakayama A, Kuwahara F, Kodama Y. An equation for thermal dispersion flux transport and its mathematical modelling for heat and fluid flow in a porous medium. Journal of Fluid Mechanics. 2006;**563**:81-96

[15] Calmidi VV, Mahajan RL. The effective thermal conductivity of high porosity fibrous metal foams. Transactions of the ASME, Journal of Heat Transfer. 1999;**121**:466-471

[16] Calmidi VV, Mahajan RL. Forced convection in high porosity metal foams. Transactions of the ASME, Journal of Heat Transfer. 2000;**122**:557-565

[17] Dukhan N. Correlations for the pressure drop for flow through metal foam. Experiments in Fluids. 2006;**41**: 665-672

[18] Kuwahara F, Yang C, Ando K, Nakayama A. Exact solutions for a thermal non-equilibrium model of fluid saturated porous media based on an effective porosity. Transactions of the ASME, Journal of Heat Transfer. 2011; **133**(11):112602

[19] Yang C, Ando K, Nakayama A. A local thermal non-equilibrium analysis of fully developed forced convective flow in a tube filled with a porous medium. Transport in Porous Media. 2011;**89**:237-249

[20] Yang C, Kuwahara F, Liu W, Nakayama A. Thermal non-equilibrium forced convective flow in an annulus filled with a porous medium. The Open Transport Phenomena Journal. 2011;**3**: 31-39

[21] Kamiuto K, Miyoshi Y, Kinoshita I, Hasegawa S. Conduction in optically thick ceramic porous media: Radiative heat transfer for the case of cordierite foam. Transactions of the JSME, Series B. 1983;**49**:2147-2153

[22] Nakayama A, Kuwahara F. A general macroscopic turbulence model for flows in packed beds, channels, pipes and rod bundles. ASME Transactions Journal of Fluids Engineering. 2008;**130**(10):101205

[23] Smirnova O, Fend T, Peter S, Schollgen D. Homogeneous and inhomogeneous model for flow and heat transfer in porous materials as high temperature solar air receiver. In: Proceedings of the COMSOL Conference; Paris. 2010. pp. 17-19

[24] Agrafiotis C, Mavroidis I, Konstandopoulos AG, Hoffschmidt B, Stobbe P, Romero M, et al. Evaluation of porous silicon carbide monolithic honeycombs as volumetric receivers/ collectors of concentrated solar radiation. Solar Energy Materials and Solar Cells. 2007;**91**:474-488

Chapter 4

Sintered Iron-Rich Glass-Ceramics and Foams Obtained in Air and Argon

Nicolai B. Jordanov, Esmat M.A. Hamzawy, Dragomir Tatchev and Alexander Karamanov

Abstract

The subsequent synthesis of sintered self-glazed glass-ceramics and/or glass-ceramic foams using metallurgical slag is the topic under discussion. The observed intensive sample expansion can be considered as an autocatalytic process related to the oxygen release due to thermal reduction of Fe_2O_3 and MnO_2 present in the slag. The sintering of the samples is studied by optical dilatometry and the foaming process by hot-stage microscopy, while the structure of the final materials is revealed by 3-D computed tomography and SEM. The phase composition of the glass-ceramic foams is analyzed by XRD. The species are characterized by moderate crystallinity, 80–85 vol.% porosity and fire resistance above 1000°C. The innovative point of this study is the synthesis in argon of sintered glass-ceramic materials where reduction is inhibited, together with a double-stage foam formation in air and argon, leading to lower working temperatures and better material characteristics.

Keywords: industrial waste, glass-ceramics, sintering, foaming

1. Introduction

The development of methods and technologies for the production of glass-ceramic materials and foams, as a particular case of glass-ceramics, has recently revealed an interesting economically grounded field for research in the synthesis of advanced materials with unique properties for application in the civil engineering and in a variety of different industries.

Many industrial streams in the metallurgy provide large amounts of waste raw materials, e.g., in the form of slag, containing the necessary composition making it suitable to be entirely recycled with just minor modifications into new glass-ceramic materials via melting and a subsequent sinter-crystallization treatment [1, 2]. This is especially appropriate when speaking of large-scale production projects.

The most significant producers of iron-rich waste are the iron and steel industries. The different final products (e.g., steel or cast iron) and the different processes used determine the variations in the waste material composition [1]. Blast furnace slag (BF) is easier to be converted to a glass and further to glass-ceramics [3] than the other slags due to the high content of silica and alumina. Other slags like the basic oxygen furnace slag (BOF) and electric arc furnace slag (EAF) are on

 IntechOpen

the contrary more enriched in iron oxides but far more poor in glass formers. They can be used after a certain composition modification [4]. Zinc hydrometallurgy turns out to be another important source of iron-rich waste raw material. Most recent research refers to iron-rich waste from the nonferrous metallurgy [5, 6].

Recent developments in the field of thermal analytical methods and instrumentation result constantly in cutting-edge machines providing faster and far more reliable laboratory measurements. In particular the high-temperature imaging systems like modern optical dilatometers (ODLT) allow in situ contactless observations of the sintering shrinkage during the synthesis of ceramics and glass-ceramics and technology development on a laboratory scale as well. Among the most significant physical parameters governing the technology of sintered glass-ceramic production are the firing temperatures, the thermal treatment rates, the thermal regimes governing the viscosity of the forming material and, as it turns out, the used atmosphere (air or inert).

An innovative part of the presented research is the synthesis in a dual air/argon environment. The authors have initiated pioneering research in this field of glass-ceramic synthesis staring in the year 2000 with some experimentation with air and inert nitrogen atmosphere [7–9] and a further development recently (started 2017) by N.B. Jordanov and A. Karamanov with argon and dual air/argon atmosphere. This development allows currently the process of synthesis of glass-ceramic samples to be separated and carried out in different stages of controlled environment, thus allowing the effect of redox determined processes to be carefully studied, understood and implemented in the design of new materials. At the same time, the crystallinity of thus obtained materials is considered higher, leading to better properties [1, 10].

The influence of the atmosphere in the particular case of foaming in inert environment of sintered iron-rich pressed powder samples however has not been examined yet and is currently the subject of profound investigations.

The overall theoretical description of effective mechanical properties and structure of cellular solid foam materials was considered and proposed first by Gibson and Ashby in 1982 [11] in terms of their famous original equations.

The theoretic development of Gibson and Ashby on the closed-cell foams however is based generally speaking on the presence of regular shape and size (e.g., cubic or hexagonal) of the closed cells determining the foams. When the case is foam of irregular structure, arbitrary shape and broad cell volume distribution, then the method of 3-D computed tomography is very helpful toward the description and characterization of cellular materials [12, 13].

In the case of cellular solid mechanics, Gibson and Ashby provide equations of the effective Young modulus of foams of closed porosity based on wide experimental data. Here the 3-D computed tomography aid is highly welcome. Actually a certain restriction here is the assumption that all closed cells are filled with a fluid. In the particular case of iron-rich glass-ceramic foams however, the inner cellular space is filled with gas. If one assumes that the pressure inside the cells is comparable to the atmospheric pressure, we can write the following equation of Young modulus [14]:

$$\frac{E^*}{E_V} = C_1\varphi^2\left(\frac{\rho^*}{\rho_V}\right) + C_2(1 - \varphi)\frac{\rho^*}{\rho_V} \quad (1)$$

where E^* is the effective modulus of the foam, E_V is the volume modulus, φ is the ratio of cell edge to the whole solid part, ρ^* is the effective density of the foam, ρ_V is the volume density, and C_1 and C_2 are geometry shape constants.

The successful synthesis in the presented investigation of glass-ceramic foams from simultaneously iron-rich and manganese-rich waste slag is crucially determined besides by the thermal treatment, by the redox couple ratio equilibria of Fe(II) ↔ Fe(III) and Mn(III) ↔ Mn(IV) [15]. The state of these redox couples is of greatest importance since it is entirely responsible for the foaming process to proceed successfully. The latter is being realized by the release of gas molecules in the bulk of the sample in the form of oxygen microbubbles determining the porosity. This is possible due to the reversible thermal partial reduction of the manganese and iron oxides of higher oxidation state into lower states, taking place during thermal foam formation. This process is vastly influenced by the environment. Recently it has been shown by the authors and other research teams as well that not only carrying out the thermal synthesis in atmospheric air but also separating the synthesis to subsequent stages in air and in inert gas environment (e.g., argon) can lead to very surprising results. Materials with different structure and properties could be thus obtained. The conditions of synthesis are also technologically favored this way using different environments.

Having already synthesized the sintered glass-ceramic material with appropriately formulated composition, one has obtained a partially crystalline and densified final product with more than satisfactory industrial features obtained at economically favorable conditions (compared to the expensive production of classic ceramic materials). Thus obtained materials from iron-rich industrial slag are ready for implementation in a number of engineering projects as well. They can be (and are actually ready to be) however further processed in terms of just a well-engineered additional thermal exposure scan at temperatures generally speaking higher than the temperature of sintering. This can be realized either isothermally or by a linear thermal scan.

The use of the novel applied method of sinter-crystallization developed experimentally in the last two decades and still being subject to theoretical progress can lead to very promising results mainly due to the relatively moderate treatment temperatures required for simultaneous sintering and crystallization of the samples. Just for the reference of the reader, the foaming process is taking place at higher temperatures after completion of the sintering and the phase formation. Glass-ceramic foams are a particular case of the general division of the so-called cellular glasses possessing high surface area, low density, low specific heat, high thermal and acoustic insulation and high chemical resistance [10]. When most of the cells are closed, the material is referred as foam.

The synthesis of sintered glass-ceramics depends on the relationship between viscous flow sintering and crystallization, while the production of glass-ceramic foams depends on the balance between apparent viscous flow sintering and gas evolution [1].

The gas release usually depends on the oxidation or decomposition reactions with the modifying compounds in glass-ceramic foaming. Oxidation reactions are associated with the release of CO_x gas from carbon-containing compounds like carbon black, graphite, silicon carbide (SiC) and organics reacting with the oxygen in the atmosphere. Typical decomposition reactions are such with carbonates or sulfates leading to the release of CO_2 or SO_x [1, 10]. A special case is when the parent glass contains intermediate oxides (called conditional glass formers) either iron oxides or manganese oxides undergoing transition from higher to lower oxidation state, connected to the release of oxygen gas. This is the actual case here which is the subject of investigation in the presented research.

So the production of glass-ceramics from iron-rich metallurgical slag can be terminated at the stage of obtaining just well-sintered glass-ceramics and extended any time further to the stage of glass-ceramic foams (in the particular case of, e.g., iron-rich parent frit).

2. Sinter crystallization and foaming

The sintering of glass-ceramics is a typical example of induced structural densification of a solid sample. The latter is provoked by volume-density variations in the material's bulk. This is the first step (besides an eventual parallel oxidation process) taking place during new glass-ceramic material formation determined by, e.g., a linear thermal scan. The degree of densification is one of the most important characteristics of sintered glass-ceramics. It is being mainly determined to a great extent by the granulometric composition, the crystallization ability (the crystallization trend) of the parent glass and the rate of heating of the compressed powder sample. At an elevated crystallization trend, the sintering could be blocked because the higher the crystallization trend, the lower the sinter ability and vice versa, respectively. It is well known that the sinterability of a glass-ceramic powder can be significantly improved by using finer glass fractions and/or higher constant linear heating rates [1, 10].

One can thus make the conclusion (and this is actually the case!) that both processes should be carefully balanced in order for a really good material with increased indicators to be successfully produced. This means that both extreme cases of a minimal or a maximal crystallization of the glass-ceramics should be avoided. The crystalline structure determines the stability and durability of the material. On the contrary, a species with higher crystallinity however cannot be sintered.

In the framework of current state of knowledge, it is assumed that both processes of sintering and crystallization are taking place in the same temperature interval. For the sake of a theoretical description however, it is accepted that the sintering stage precedes the crystallization stage. In fact the sintering stops after formation of a critical percentage of crystal phases.

Of crucial importance here is the apparent viscosity of the glass-ceramics. Its value should be maintained in a range, such that the expansion of the structure is possible and the formation and the propagation of open porosity are temporarily unavailable.

The foaming in the studied case is determined by the formation and distribution of a closed porosity population in the bulk of the material due to the release of gas molecules (most often oxygen) during the high-temperature partial reduction of certain oxides from higher to lower oxidation state. These are most often the iron [16] and the manganese [15] oxides. The high-temperature interval of foaming indicates that the mechanism of gas formation is directly related to the oxygen release as a result of the reversible partial Fe(III) and Mn(IV) reduction [15–19]. Moreover when the oxides of the iron and the manganese are naturally present in a sufficient amount in the parent glass frit, this is the most favorable case because there is no need for the process of foaming to be artificially and additionally catalyzed. An inorganic material is thus being formed by autocatalytic foaming.

3. Experimental

For the purpose of current experimentation with sintered materials and foams, a slag from the iron and steel company Helwan, Cairo Governorate, Egypt (with a slag production capacity of 30 kt p.a.), was used for the synthesis of the investigated samples.

This slag is relatively poor in glass formers (SiO_2 and Al_2O_3); that's why it had to be enriched in silica by mixing 70 wt.% slag and 30 wt.% industrial sand. The parent batch of 150 g was brought to melting in corundum crucibles using an electric furnace. After an exposure for 2 hours at a temperature of 1450°C, the resulting homogenous melt was quenched in water, and a dark brown-colored glass frit was obtained. Thus obtained glass frit was crushed, grinded several times in a planetary mill FRITSCH (Germany) for 10 minutes and sieved below 75 μm with a digitally programmed sieving machine CISA (Spain).

The investigations were performed by a thermal-optical measuring and imaging system with an ESS HSM-1400 MISURA (Italy). This instrument combines two techniques: a high-precision, high-resolution horizontal contactless optical dilatometer and a hot-stage microscope (HSM). This is an established laboratory method in recent years, since it turns out to be reliable and fast and is used already by many research groups worldwide [2, 20].

The sintering behavior of all glass-ceramic samples was investigated by means of ODLT, a method which allows measurements with very high precision. This is mainly due to the absence of a mechanical push rod in the system, as it is the case with classical contact dilatometer devices. A monochrome optical arrangement employing two video cameras providing high magnification and high resolution is used instead. Typical measurements were performed with holding times of 30 minutes at 950°C in air or argon.

To the sintered glass powder, a small amount of 7 wt.% polyvinyl alcohol (PVA) aqueous solution was added to form a granular mass, which then is mechanically homogenized and placed in a matrix of 50 × 5 × 3 mm. Multiple samples of equal forming are prepared by stuffing loosely the material in a pressing matrix. Then by applying uniaxial pressure at 40 MPa in a hydraulic pressing machine NANNETTI (Italy), samples of almost perfectly equal dimensions and densities, with an increased green strength and a decreased porosity, are produced.

Subsequently a burnout step at 270°C is required to be carried out before proceeding further with the heat treatment processes in order to remove the binding agent (e.g., PVA). The burnout can be performed in ODLT separately or as an initial programmed step preceding the oxidation, sintering and foaming steps.

The studied thermal behavior in the range up to 1300°C in air and argon of the sintered samples was studied and analyzed by the optical HSM method. As far as the samples used for ODLT measurements have a standard parallelepiped shape (as described above), the samples used for HSM measurements are not the same but represent upright standard cylinders instead.

In addition the foaming trends of the samples were examined by carrying out measurements both isothermally with holding times of 30 minutes (e.g., at 950°C for simultaneous oxidation and sintering and at 1100°C for foaming initiation) and non-isothermally as well (by using constant linear thermal scans). Typically such thermal scans were used with heating rates of 20°C min^{-1}.

During all measurements the ODLT/HSM instrument was mounted in a closed aluminum box (developed and manufactured at the mechanical workshop of IPC-BAS Sofia), allowing measurements in a controlled environment. The box can be purged with dry argon gas on demand. The latter application allows continuous maintenance of an overpressure of ~10 mbar argon to be kept over the entire synthesis and optical measurement of the thermal variation of the structure of the sample. This is achieved by the use of a fine-graded rotameter, Yokogawa (Japan), for manual control of the gas flux in the sample chamber and a high-precision digital manometer with a ceramic membrane, Profimess (Germany), for overall gas pressure monitoring.

According to the current state of knowledge of the authors, such a laboratory experimental setup—a combined ODLT/HSM mounted in a closed vessel for examinations of the sintering and foaming behavior of iron-rich glass-ceramics in different atmospheres—is unique in Southeastern Europe and even was probably used for the first time in this respect here.

Scanning electron microscopy (SEM) was used to analyze the structure of the sintered glass-ceramics by taking pictures of both fractures and the surfaces of the samples. A JEOL 6390 (Japan) instrument was used. To provide electron conductivity, all samples were metalized with gold by vapor deposition technique.

3-D computed micro-tomography was used for entire bulk scanning of the foam glass-ceramic species. The tomographic measurements were carried out with an X-ray micro-tomograph, Bruker SKYSCAN 1272 (Germany), which uses a white beam with cone geometry. The following setup conditions were applied: X-ray tube voltage 70 kV, current 142 mA and 0.11 mm Cu filter. The voxel (3-D pixel) size was 1 μm and the optical magnification was 7.4. A typical 360° scan took 21 hours and 27 minutes. Reconstruction of the 3-D images was done with the commercial software InstaRecon.

The phase composition of the sintered glass-ceramic foams was determined by X-ray diffraction spectroscopy (XRD) using a Panalytical Empyrean (USA) spectrometer.

4. Summary of results and discussion

As we have already noted, the innovative point of current research is, besides the use of the applied method of sinter-crystallization toward the production of sintered glass-ceramics and/or glass-ceramic foams, the use of a controlled environment during the synthesis of sintered glass-ceramics and in particular extended to the design and production of glass-ceramic foams.

In **Figure 1**, a comparative graphical representation is presented, together with images at respective temperatures of interest, of the shape alteration during real-time in situ filming of two samples being synthesized in air or in argon atmosphere.

Up to temperatures of about 1050°C, the thermal behavior of both species is quite comparable. Both samples undergo similar densification during the interval of low temperature sintering. At higher temperatures however, iron-rich glass-ceramic materials heated up in different (air or inert) atmosphere reveal a completely different thermal behavior compared to each other when they are subject to a subsequent thermal treatment.

In the case of an argon-sintered sample, further heating of the material leads to sample melting (to, e.g., a hemispheric spill at 1155°C). However, the sample sintered in air starts expanding its volume during heat treatment above 1050°C, and at the same reference temperature of 1155°C, it reveals a maximal value of the structural expansion determined by gas evolution—the foaming in the entire bulk of the newly formed material. This difference in the thermal behavior of both samples is mainly due to differences in the respective material's viscosity values.

The phenomena described above are entirely determined by the state of the redox couple equilibrium Fe(II) ↔ Fe(III) in a sense that an inert environment (e.g., argon) is going to maintain the equilibrium drawn at a maximal extent to the left (oxidation is inhibited [9]), while the air atmosphere is going to keep the ratio Fe(II)/Fe(III) at a minimal value, i.e., the equilibrium is drawn to the right.

In **Figure 2a**, the sintering curves are presented during an isothermal scan and the respective low temperature behavior of the glass-ceramic samples in air and argon. One can clearly and unambiguously note the effect of the environment on sintering: the onset of the sintering process in argon is shifted to lower temperatures. It starts earlier, and the degree of structural densification is to a certain extent higher than the one in air (14 vs. 12% shrinkage, c.f. again **Figure 2a**). The shaded area in **Figure 2a** represents actually the area of temperatures where the glass transition point, T_g, which is the most important physical characteristic of a glass, is to be observed.

In **Figure 2a**, the shaded area reveals in fact, as it is obvious in **Figure 2b**, that an ODLT measurement represents certainly a method for truly reliable measurements

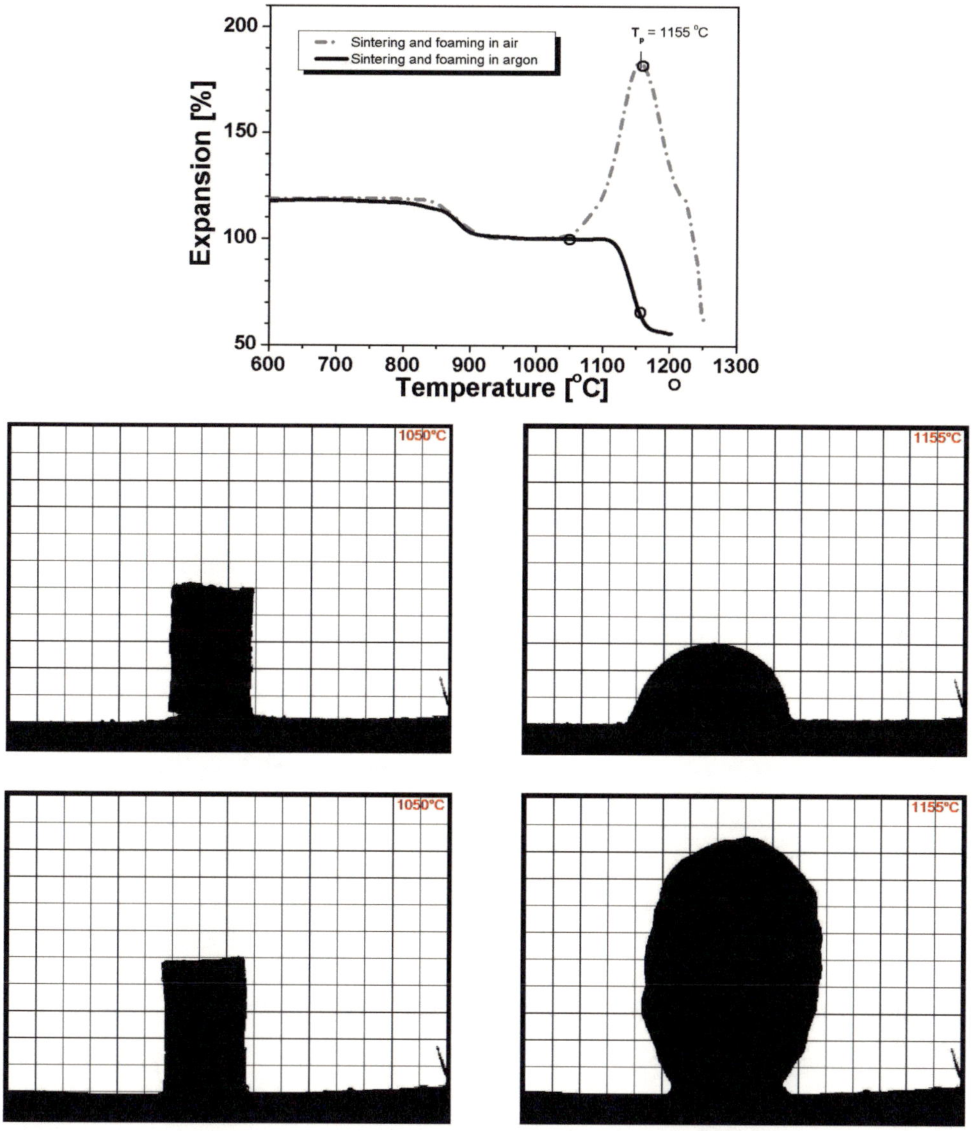

Figure 1.
Typical HSM curves of the thermal behavior of pressed iron-rich glass-ceramic samples in air and argon.

of the temperature of vitrification, T_g, of pressed powder objects as well. It is evident that the glass transition point of the sample sintered in argon is significantly diminished and the difference in T_g between glass-ceramics synthesized in air and argon amounts to ΔT = 20°C.

One can thus summarize that utilizing the ODLT technique for the sake of synthesis of well-sintered glass-ceramic materials is a good approach. It also reveals the possibility for reliable measurements of T_g, due to its high precision as a result of lack of mechanical parts, and for exact tuning of the appropriate firing regimes.

XRD spectra of the glass-ceramic foams obtained in air and argon environment are presented in **Figure 3**. The phase analysis of the samples results in the detection of pyroxene crystal phases. The crystallinity of the samples is between 30 and 40% as referred also by Strnad [21]. The crystallinity in argon however is higher than that in air. A shift of the peak positions in air to lower theta angles has been observed as well.

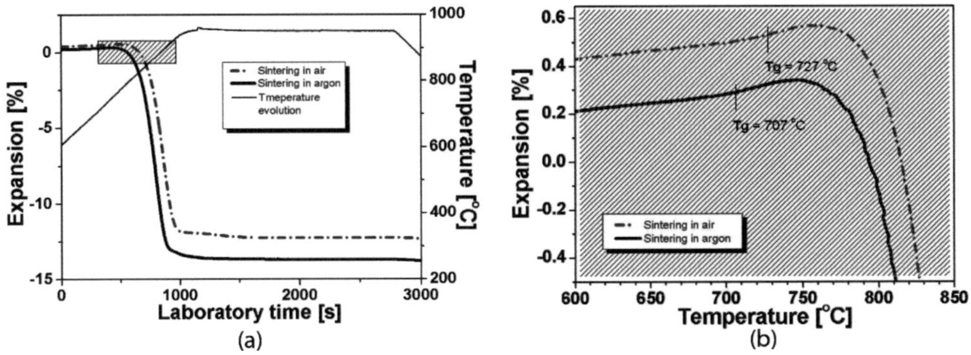

Figure 2.
(a) Sintering curves measured by ODLT in air and argon. (b) Sintering curves; zoom-in of the shaded area of (a). The glass transition temperature (Tg) is unambiguously to be recognized in both curves.

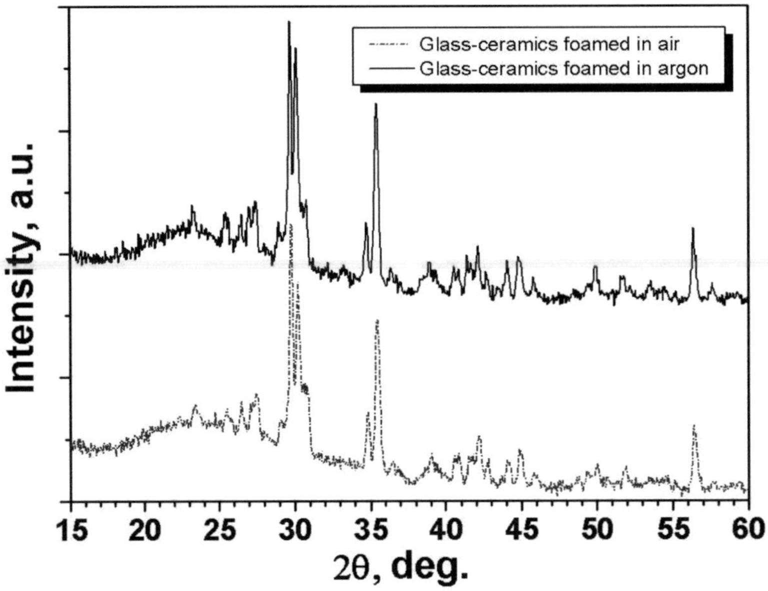

Figure 3.
XRD spectra of glass-ceramic foams synthesized in different environment.

The two features described above confirm that the absence of oxidation in argon environment leads unambiguously to higher crystallinity and to some differences in the chemical composition of the pyroxene phases. These differences can hypothetically be due to the facilitated entering of the Fe(II) and Mn(III) ions in the pyroxene structures. Also a change in the lattice interplanar distance is present.

In order for the microstructure of the sintered glass-ceramic species subject to current investigation to be studied, a series of scanning electron microscope images of two sintered glass-ceramic samples are shown in **Figure 4**. Images have been taken from the surface and from a fracture of both species. **Figure 4a** and **c** represents photos of the surface of the air-sintered sample; **Figure 4b** and **d** is from the surface of the argon-sintered sample; **Figure 4e**, **g** and **h** is images from a fracture of the air-sintered sample; and **Figure 4f** is a picture of a fracture of the argon-sintered sample.

Despite the well overall sintering in both atmospheres, it has been found that in argon environment an even better sintering of the sample than that in air is present.

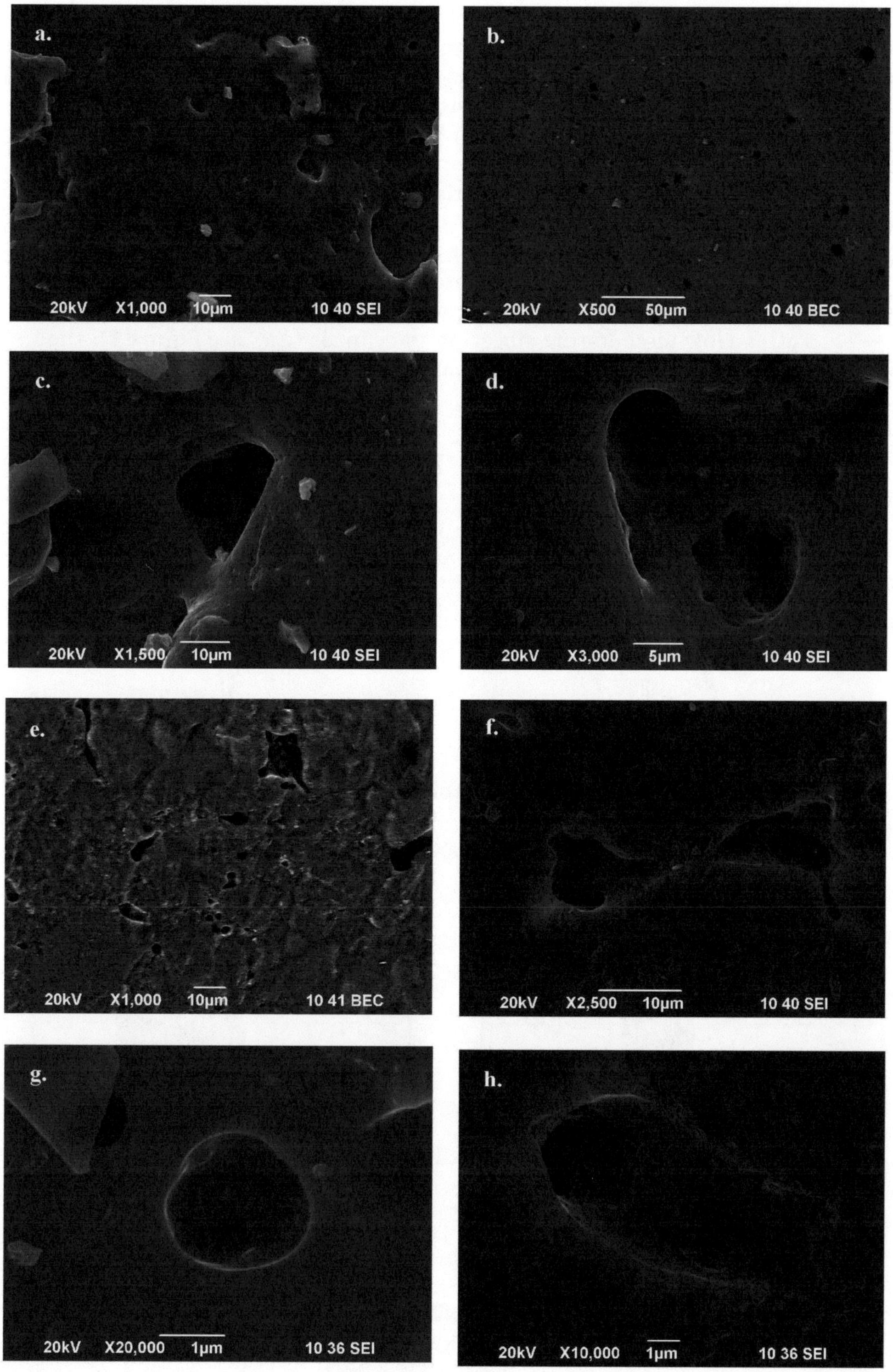

Figure 4.
SEM images from a fracture and the surface of two sintered glass-ceramic foam samples in air and argon: 4a and 4c: surface of the air sintered sample; 4e, 4g, 4h: fracture of the air sintered sample. 4b and 4d: surface of the argon sintered sample; 4f: fracture of the argon sintered sample.

An interesting observation is that on the surface of the argon-sintered glass-ceramics (**Figure 4b**), the pores reveal a concave morphology. This might be an indication for the formation of these pores in the interval of softening which initiates together with the beginning of the process of high-temperature reduction as well. In both cases of sintering, a large population of small spherical holes on the surface is present, which is almost a certain indication for gas release.

In both cases of sintering, a non-spherical, sharp-edge intergranular porosity is predominant.

The proposed possibility for a selective environment of synthesis of sintered glass-ceramics and glass-ceramic foams provides the option of the synthesis to be carried out and thus to be experimented by using different stages.

Thus after a low temperature sintering stage in air (e.g., see **Figure 1**), the process of foaming of a pressed glass sample can be carried out in a second stage

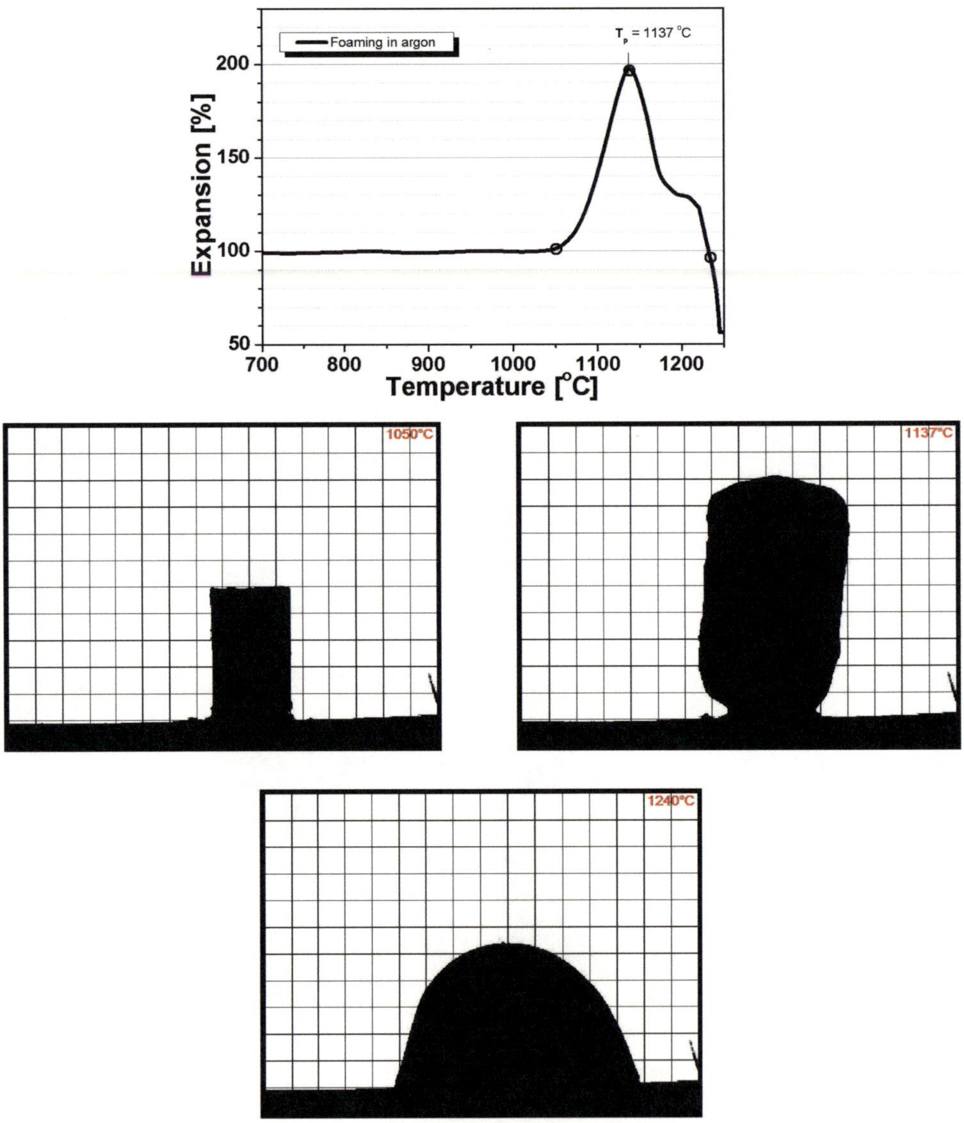

Figure 5.
Measurements of the thermal shape alteration behavior during foaming in argon by HSM; snapshots of the sample at a respective characteristic temperature.

entirely in argon environment. The process of foaming of this sample traced by a HSM measurement is shown in **Figure 5**. The snapshots in **Figure 5** during the run of a thermal scan represent the evolution of structural alteration at characteristic temperatures.

In order for the effect of the atmosphere on the foaming of sintered glass-ceramics to be carefully analyzed, a graphical plot of the structural evolution during foaming in different atmospheres is given in **Figure 6**. The plot in the expansion-time domain of foaming in air and argon during an isothermal scan at a temperature of 1100°C with 30 minutes holding time (**Figure 6**) provides a clear and detailed picture on the process under consideration.

The expansion of the structure due to the release of oxygen gas and formation of the closed-cell system during high-temperature reduction initiates in both environments (air and inert) in a similar way and almost at the same time as well. The process of foaming in argon however turns to be more effective. The bloating of the material reaches a maximal value of nearly 200%, considerably earlier than the maximal foaming in air (up to 190%); then after a slight volume shrinkage, a stable material with 180% expanded structure is being formed and maintained in the course of the working isotherm.

The entire foam material formation (i.e., the formation of closed porosity population) proceeds relatively quickly with the programmed isotherm in **Figure 6**. The foaming in argon takes ~5 minutes to complete followed by a small shrinkage and a structural stabilization. The foaming in air takes longer time to complete than argon (15 minutes) and reaches stabilization again. The foam material formation in air and argon atmospheres results in obtaining a new material characterized in both cases generally with fire resistance properties at temperatures of 1100°C.

For the sake of investigation of the entire bulk structure of the newly formed foam material, 3-D X-ray tomographic analysis has been used. In **Figure 7a** (left), a false-color 3-D reconstruction of the surface is presented and the volume by tomographic scanning of an iron-rich glass-ceramic foam sample synthesized in air. In **Figure 7a** (right), a selection of three cross sections of the bulk of the same sample is presented. In **Figure 7b** (left), similar 3-D false-color surface-volume reconstruction of a glass-ceramic foam sample synthesized in argon is presented. In **Figure 7b** (right), cross-section slices of the material's bulk are shown, respectively.

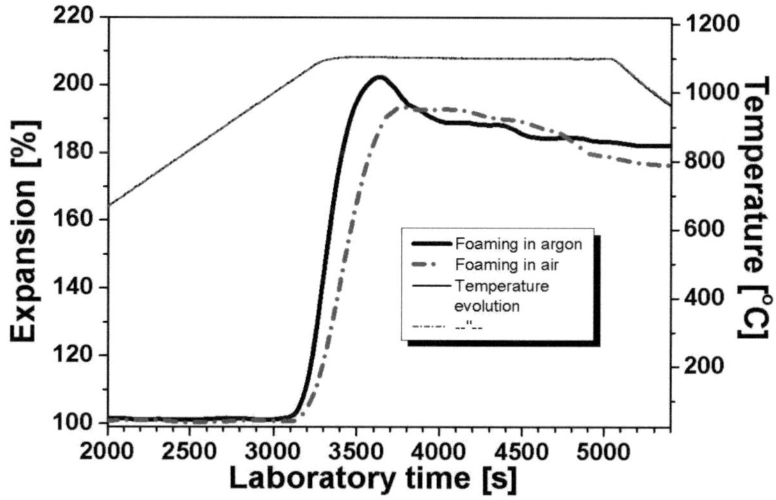

Figure 6.
Temporal evolution of foaming curves in air and argon of sintered glass-ceramic samples.

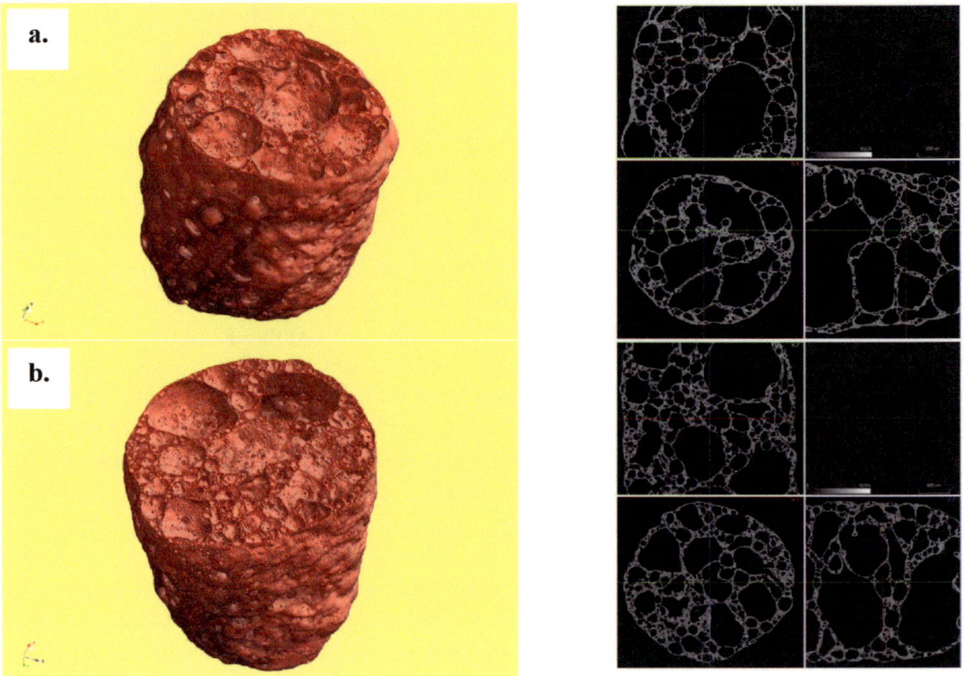

Figure 7.
X-ray computed tomography, (a) 3-D reconstruction of the surface and the bulk of a glass-ceramic foam synthesized in air (left); volume cross sections (right), and (b) 3-D reconstruction of the surface and the bulk of a glass-ceramic foam synthesized in argon (left); volume cross sections (right).

From visual observations and the performed analysis, one can clearly note that the porosity in the bulk of both samples is predominantly of closed type and amounts to about 80–85% in both species. It is also evident that the walls of the samples are abundant of pores. Another interesting feature is that the closed cells of the species foamed in air are larger than the cells in the volume of the sample foamed in argon. Moreover the thickness of the walls in the sample synthesized in air is on average lower than the wall thickness in the glass-ceramic foam material obtained in argon environment.

5. Conclusions

In current investigation the authors have shown the possibility for carrying out successful synthesis depending on the temperature and the applied atmosphere of well-sintered glass-ceramics and/or glass-ceramic foams.

It was shown that synthesis in inert environment leads to the production of materials with higher degree of sintering.

If properly engineered, the practice and experience show that a glassy-crystalline material with low density and high porosity of about 80% can be obtained, which can be considered as more than a satisfactory result [1, 22]. In addition thus obtained foams are characterized by fire resistance features up to 1100°C as well.

By applying a double-stage heat treatment, first stage, low temperature sintering at 950°C, and second stage, foaming at 1100°C, foam materials with improved properties and differences in the structure have been successfully obtained.

The SEM images from the fractures of both glass-ceramic foam species reveal a well-sintered material. They show a good degree of sintering and total porosity below 10–15%.

Inorganic materials obtained by thermal foaming like these subjects of current investigation can be used as modern building materials, as construction materials and as insulation materials in various fields and industries, e.g., as panels or arbitrary shaped.

Inorganic glassy-crystalline foam materials can be considered generally speaking as low-cost thermal insulating, soundproof and fire-resisting low-weight materials.

Acknowledgements

N.B. Jordanov gratefully acknowledges the financial support of the National Program Young Scientists and Post-docs of the Ministry of Education and Science of Republic of Bulgaria under the project "Sintered self-glazed glass-ceramics and foams from iron-rich industrial wastes." D. Tatchev and A. Karamanov express their gratitude for the support of Project BG05M2OP001-1.002-0019: "Clean technologies for sustainable environment—waters, waste, energy for circular economy." All authors are thankful for the financial support for realization of the closed experimental setup for measurements in a dual environment of the Bulgarian Science Fund (FNI) under Project DN 19/7 "Theory and application of sinter-crystallization."

Conflict of interest

The authors declare no conflict of interest.

Author details

Nicolai B. Jordanov[1], Esmat M.A. Hamzawy[2], Dragomir Tatchev[1] and Alexander Karamanov[1]*

1 Institute for Physical Chemistry, Bulgarian Academy of Sciences, Sofia, Bulgaria

2 Glass Research Department, National Research Centre, Cairo, Egypt

*Address all correspondence to: karama@ipc.bas.bg

References

[1] Rincón A, Marangoni M, Cetin S, Bernardo E. Recycling of inorganic waste in monolithic and cellular glass-based materials for structural and functional applications. Journal of Chemical Technology & Biotechnology. 2016;**91**:1946-1961. DOI: 10.1002/jctb.4982

[2] Fiocco L, Elsayed H, Daguano JK, Soares VO, Bernardo E. Silicone resins mixed with active oxide fillers and Ca–Mg silicate glass as alternative/integrative precursors for wollastonite-diopside glass-ceramic foams. Journal of Non-Crystalline Solids. 2015;**416**:44-49. DOI: 10.1016/j.jnoncrysol.2015.03.001

[3] Fredericci C, Zanotto ED, Ziemath EC. Crystallization mechanism and properties of a blast furnace slag glass. Journal of Non-Crystalline Solids. 2000;**273**:64-75. DOI: 10.1016/S0022-3093(00)00145-9

[4] Ferreira E, Zanotto ED, Scudeller L. Glass and glass-ceramic from basic oxygen furnace (BOF) slag. Glass Science and Technology. 2002;**75**:75-86

[5] Karamanov A, Aloisi M, Pelino M. Vitrification of copper flotation waste. Journal of Hazardous Materials. 2007;**140**:333-339. DOI: 10.1016/j.jhazmat.2006.09.040

[6] Sarraffi A, Rahmati B, Hassani HR, Shirazzi HHA. Recovery of copper from reverberatory furnace slag by flotation. Minerals Engineering. 2004;**17**:457-459. DOI: 10.1016/j.mineng.2003.10.018

[7] Karamanov A, Taglieri G, Pelino M. Sintering behavior and properties of iron-rich glass-ceramics. Journal of the American Ceramic Society. 2004;**87**(7):1354-1357. DOI: 10.1111/j.1551-2916.2004.01571.x

[8] Karamanov A, Pisciella P, Pelino M. The crystallisation kinetics of iron rich glasses in different atmospheres. Journal of the European Ceramic Society. 2000;**20**:2233-2237. DOI: 10.1016/S0955-2219(00)00077-7

[9] Karamanov A, Taglieri G, Pelino M. Sintering in nitrogen atmosphere of iron-rich glass-ceramics. Journal of the American Ceramic Society. 2004;**87**(7):1354-1357. DOI: 10.1111/j.1151-2916.2004.tb07734.x

[10] Scarinci G, Brusatin G, Bernardo E. Glass foams. In: Scheffler M, Colombo P, editors. Cellular Ceramics: Structure, Manufacturing, Properties and Applications. Weinheim: Wiley-VCH; 2005. pp. 158-176. DOI: 10.1002/3527606696

[11] Gibso LJ, Ashby MF. Cellular Solids: Structures and Properties. 2nd ed. Cambridge: Press Syndicate of University of Cambridge; 1997. DOI: 10.1017/CBO9781139878326

[12] Elmoutaouakkil A, Salvo L, Maire E, Peix G. 2D and 3D characterization of metal foams using X-ray tomography. Advanced Engineering Materials. 2002;**4**(10):803-807. DOI: 10.1002/1527-2648(20021014)4:10<803:AID-ADEM803>3.0.CO;2-D

[13] Stock SR. Recent advances in X-ray microtomography applied to materials. International Materials Review. 2008;**53**(3):129-181. DOI: 10.1179/174328008X277803

[14] Ma Y, Su X, Pyrz R, Rauhe JC. A novel theory of effective mechanical properties of closed-cell foam materials. Acta Mechanica Sinica. 2013;**26**(6):559-569. DOI: 10.1016/S0894-9166(14)60001-X

[15] Petersen RR, Koenig J, Yue Y. The mechanism of foaming and thermal conductivity of glasses foamed with MnO_2. Journal of Non-Crystalline

Solids. 2015;**425**:74-82. DOI: 10.1016/j.jnoncrysol.2015.05.030

[16] Sandrolini F, Palmonary P. Role of iron oxides in the bloating of vitrified ceramic materials. Transactions and Journal of the British Ceramic Society. 1976;**75**:25-32

[17] Appendino P, Ferraris M, Matekovits I, Salvo M. Production of glass ceramic bodies from the bottom ashes of municipal solid waste incinerators. Journal of the European Ceramic Society. 2004;**24**:803-810. DOI: 10.1016/S0955-2219(03)00264-4

[18] García-Ten J, Saburit A, Orts MJ, Bernardo E, Colombo P. Glass foams from oxidation/reduction reactions using SiC, Si_3N_4 and AlN powders. Glass Technology. 2011;**52**:103-110

[19] Chinnam RK, Francis AA, Will J, Bernardo E, Boccaccini AR. Review. Functional glasses and glass-ceramics derived from iron rich waste and combination of industrial residues. Journal of Non-Crystalline Solids. 2013;**36**:563-574. DOI: 10.1016/j.jnoncrysol.2012.12.006

[20] Karamanov A, Pelino M. Crystallization phenomena in iron rich glasses. Journal of Non-Crystalline Solids. 2001;**281**(1-3):139-151. DOI: 10.1016/S0022-3093(00)00436-1

[21] Strnad Z. Glass-Ceramic Materials: Liquid Phase Separation, Nucleation and Crystallization in Glasses. Amsterdam: Elsevier; 1986

[22] Sasmal N, Garai M, Karmakar B. Preparation and characterization of novel foamed porous glass-ceramics. Materials Characterization. 2015;**103**:90-100. DOI: 10.1016/j.matchar.2015.03.007

Chapter 5

Design and Development of Manufacturing System Design for Producing Metallic Foam

Shyam Sunder Sharma

Abstract

This chapter involves the development of a new metal foam composite material using casting techniques. This work included the design and the development of a process to produce the metal foam and control it. The materials used to produce the foam consisted of aluminum. Using prefabricated hollow spheres assures a uniform pore size and cell wall thickness. Casting a metal into the interstitial space provides a solid medium to add a structural support to the foam. The experimental setup was designed and fabricated. By the use of the developed setup, aluminum metallic foam was developed from the raw aluminum. Some characteristics were studied to verify the findings with the published data.

Keywords: metallic foam, pores, honeycomb, foaming

1. Introduction

The development of the metal foam starts in the late 1940s. Many patents were registered in the 1950s and 1970s. The popularity of metal foam arises in the 1980s, and the old techniques are modified and re-established. Researchers are trying to mechanize the foaming method to minimize the production cost and increase the productivity [1–3]. For the development of complex shapes and large quantities, powder metallurgy and casting techniques were used. For refractory metals electro-deposition method was used. The metal foam has a better strength-to-weight ratio, higher stiffness, increased energy absorption and high temperature tolerance. It gives higher stiffness than a solid metal [4, 5]. The metal foam can be classified according to the foam structure. It has an open structure, close structure and combination of open and closed structures [6, 7]. The properties of metal foam are lightweight, energy absorption, high stiffness and higher compression strength [8, 9].

There are various methods for developing metallic foam. These production methods are classified according to the state of the material and process followed. Metallic foam was categorized into four groups as given in **Figure 1**:

i. Casting method (liquid metal): In this method a liquid metal is poured into the mold, and a foaming agent is added which creates a metal foam around the solid filler metal.

ii. Powder metallurgy (solid metal in powdered form): In this process metals are converted into a powdered form, and then a foaming agent is added

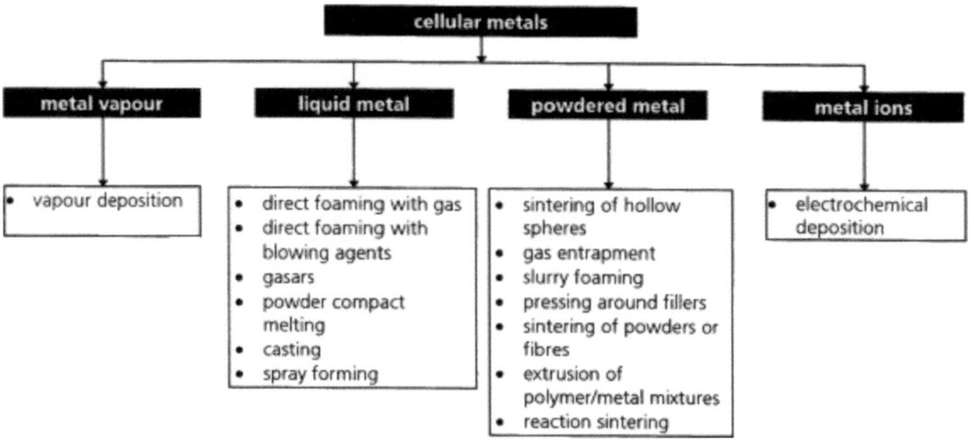

Figure 1.
Production methods of metallic foam [10].

with the metal powder; after that the blend is compressed to yield a dense. In this method, compaction can be performed by any technique that ensures that the foaming agent is embedded in the metal matrix without any residual open porosity.

iii. Metal vapor: A metal foam is commonly produced by injection of foaming gas into the liquid metal or by adding a foaming agent into the metal powder. In this method melts can be foamed by creating gas bubbles in the material. Normally, bubbles in the molten metal are highly buoyant in the high-density liquid and rise quickly to the surface.

iv. Metal ion solution: The alkali metal ions in the solution electrostatically adsorb onto the many bubbles by bubbling ionic surfactant solution, and they transfer to the air/water interface. These bubbles change to the foam state that is removed using a long glass tube.

2. Design and fabrication

For producing metallic foam liquid state, a casting processing technique was used. In this process calcium carbonate was added into the liquid aluminum and was stirred. When aluminum liquid was stirred, calcium carbonate released gases, and these gases are entrapped in aluminum and created porosity. For producing metallic foam, a setup was required, and this was fabricated in the production engineering department. For design and fabrication, the resources which are available in the department are used. This was the first setup fabricated for producing metallic foam. According to the resources, different elements have been designed and fabricated.

For fabricated setup different elements were required, and these were designed according to the requirement. Some elements are available in the lab and they are directly taken according to the requirement, and others were fabricated. Fabricated setup resources were used that were available in lab, because this was the first setup and after the successful of that setup go for the standardization. The process sheets of fabrication are given in the appendix. The design given in **Figure 2** was taken as the reference design. The elements that were required are given below:

1. Induction motor.

2. Shaft.

3. Selection of bearing and bearing socket.

4. Driver and driven pulley.

5. Milling socket.

2.1 Specification of induction motor

An induction motor was used for a rotate stirrer that was placed in the end of the shaft. When the setup will be standardized, a standard motor will be used. The speed of the motor was constant. Driver and driven pulleys were connected by a rope. The specifications of the motor are given below:

1. Power = 1 HP.

2. rpm = 1380.

3. Efficiency = 72%.

4. Machine no. = 88A615–204.

5. Diameter of the shaft = 19.19 mm.

2.2 Design of the shaft

A shaft was a rotating element which is used for a rotate stirrer and produces metallic foam.

2.2.1 Calculation

Material of shaft = Mild steel, it has tough material and it was absorbing vibration that was caused by the motor.

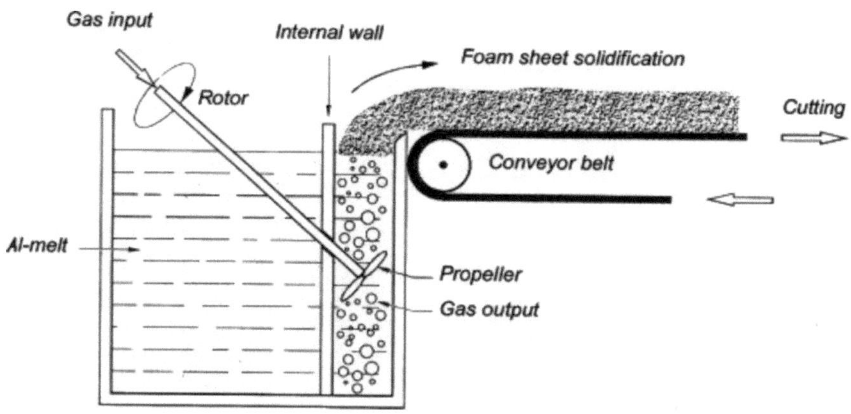

Figure 2.
Basic design for producing foam [10].

Length of the shaft = 315 mm.
Diameter of the shaft from top 225 mm = 20 mm.
Diameter of the shaft at lower 80 mm length = 19.98 mm.
Dimension of thread = 10 × 1.5 mm.

Considering the load at the top, i.e. load of the pulley mounted on the shaft.

$$\text{Slenderness ratio of the shaft} = \frac{L}{K} = \frac{315}{10} = 31.5 \tag{1}$$

Thus, the shaft is a medium type (**Figure 3**).

2.2.2 Rankine formula for calculating the crippling load

$$W_{cr} = \frac{\sigma_c \times A}{1 + \frac{\sigma_c \times A \times L^2}{\pi^2 \times E \times I}} \tag{2}$$

where W_{cr} = crippling load, σ_c = crushing stress or yield stress, A = cross-sectional area, L = length of the shaft, E = modulus of elasticity, I = moment of inertia.

$$W_{cr} = \frac{\sigma_c \times A}{1 + a\left(\frac{L}{K}\right)^2} \tag{3}$$

$$a = \text{Rankine's constant} = \frac{\sigma_c}{\pi^2 E} \tag{4}$$

$$a = \frac{1}{7500}$$

For one end fixed and another end free.

$$L = 2L = 2 \times 315 = 630 \text{ mm}$$

$$K = 10 \text{ mm}$$

$$\frac{L}{K} = \frac{630}{10} = 63$$

$$W_{cr} = \frac{320 \times \frac{\pi}{4}(20)^2}{1 + \frac{1}{7500}(63)^2}$$

$$\sigma_c = 320.$$

$$W_{cr} = 65.707 \text{ N}$$

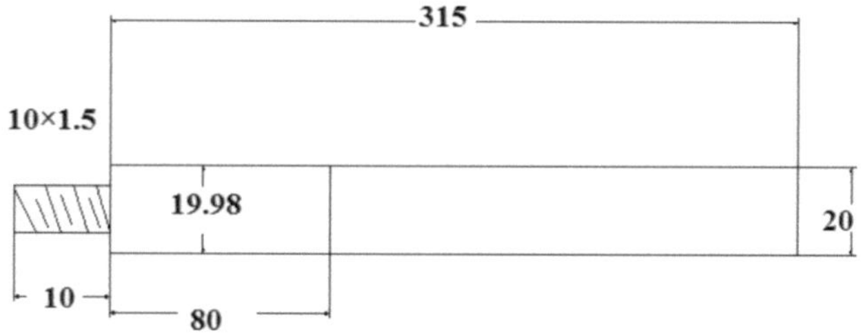

Figure 3.
Shaft.

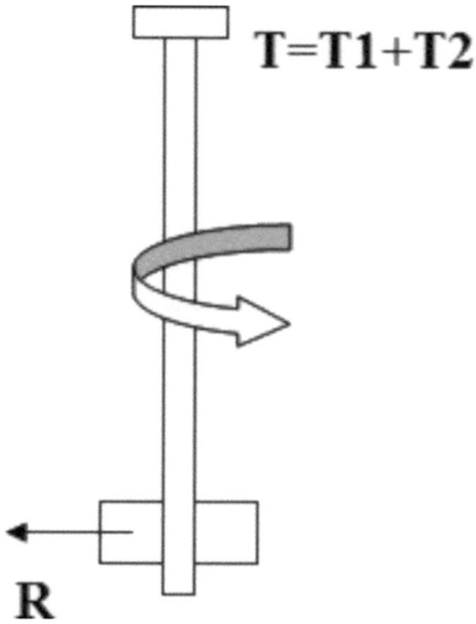

Figure 4.
Shaft in vertical position.

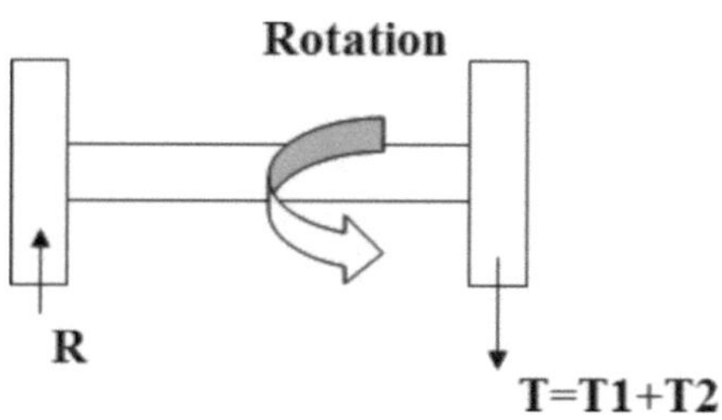

Figure 5.
Shaft in horizontal position.

2.2.3 Considering the weight of pulley

Figure 4 shows the vertical position of the shaft, and **Figure 5** shows the horizontal position of the shaft. In **Figure 6** power is transmitted from the driver pulley to driven pulley. For designing a shaft and pulley, the following calculations are required.

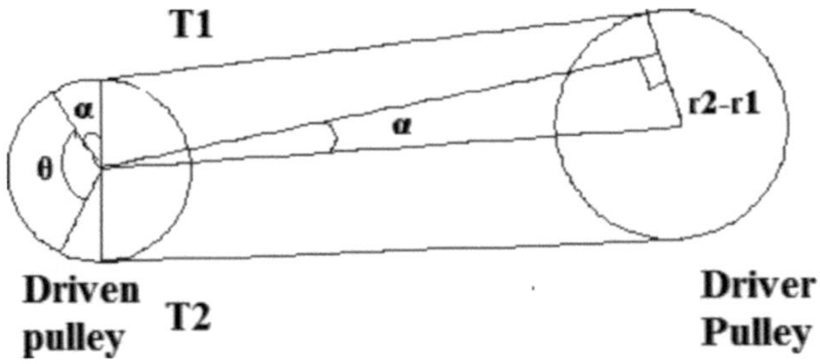

Figure 6.
Power transmitted by rope.

$$\text{weight of the pulley} = 400 \text{ g}$$
$$= \frac{400 \times 9.81}{1000}$$
$$W_p = 3.9 \text{ N}$$

The shaft bears more load; hence, the design is safe.

2.2.3.1 Power of motor

$$P = \frac{2\pi N T_0}{60} \quad (5)$$

T_0 = Torque.

$746 = \frac{2\times\pi\times1380\times T_0\times.72}{60}$.
T_0 = 7.18 N mm.
T = (T1 − T2) × 44.

$$\frac{T1}{T2} = e^{\mu\theta} \quad (6)$$

T1 = Tension in the tight side.
T2 = Tension in the slack side.
μ = Coefficient of friction.
The distance between the driven and driven pulley is 330 mm:
θ = 180-2α.
θ = Angle of contact.

$$\text{Sin}\alpha = \frac{r_2 - r_1}{330} = \frac{60 - 44}{330} = 0.484 \quad (7)$$

r_1 & r_2 = radii of the driving and driven pulley.
α = 2.778°.
θ = 180–5.55.
θ = 174.44° = 3.03 rad.
μ = 0.25.
$\frac{T1}{T2} = e^{0.25\times3.02}$
T1 = 2.12 T2.
(2.12–1)T2 × 44 = 7.18.
T2 = 0.159 N mm.

T1 = 0.33 N mm.
R = T1 + T2.
R = 0.489 N mm.
$M_b = T \times L = 0.489 \times 315 = 154.03$ N mm.
$M_t = T_O = 7.18$ N mm.

Load is steady hence.
$K_m = 1.5$.
$K_t = 1.0$.
K_m = Combined shock and fatigue factor for bending.
K_t = Combined shock and fatigue factor for torsion.

2.2.4 Equivalent bending moment

$$M_e = \frac{1}{2}\left[K_m \times M_b + \sqrt{(K_m \times M_b)^2 + (K_t \times T)^2}\right] \tag{8}$$

$$M_e = \frac{1}{2}\left[1.5 \times 154.035 + \sqrt{(1.5 \times 154.035)^2 + (7.18)^2}\right]$$

$$M_e = 231.108 \text{ N mm}.$$

Determine the equivalent bending moment:

$$231.108 = \frac{\pi}{32}\sigma_b \times d^3.$$

$$\sigma_b = \frac{231.108 \times 32}{\pi \times (20)^2}.$$

$$\sigma_b = 0.294 \text{ N/mm}^2.$$

This was very less than the allowable tensile or compressive stress. Hence the design was safe.

2.3 Selection of bearing

In order to select a most suitable ball bearing, first, the basic dynamic load was calculated. It was multiplied by the service factor to get the basic dynamic load capacity. After finding the basic dynamic load capacity, the selection of bearing was made from the catalog of a manufacturer (**Figure** 7).

2.3.1 Striebeck formula for the strength of a single ball in compression

$$F_e = Kd^2 \tag{9}$$

K = 7200.
d = diameter of ball.
$F_e = 7200 \times 7^2$.
$F_e = 352{,}800$ N/mm^2.

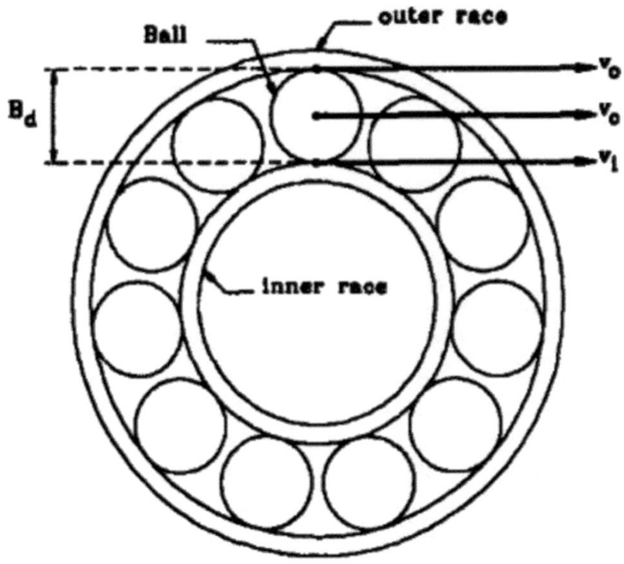

Figure 7.
Axial view of ball bearing.

2.3.2 The maximum load per ball

$$F_c = \frac{4.37}{n} \times C \tag{10}$$

$$C = \frac{F_c \times n}{4.37} = \frac{Knd^2}{4.37} \tag{11}$$

n = no. of balls.
C = capacity of the bearing.

$$C = \frac{7200 \times 8 \times .007^2}{4.37}.$$

$$C = 6.45\ \text{kg}.$$

$$F_c = \frac{4.37}{8} \times 6.45$$

$$\text{Torque} = \frac{9550 \times \text{H.P}}{N} = \frac{9550 \times .75}{1380}.$$

T = 5.19 N m.
where HP in kW, N in rpm.

2.3.3 Basic dynamic load of bearing

$$C_d = F_c \times (n \cos\alpha)^{0.7} Z^{2/3} D^{1.8} \tag{12}$$

$$= 3.46 \times (8 \cos 15)^{0.7} \times 47^{1.8}.$$

$C_d = 14807.3$ N.
$C_d = 14.8$ kN.

α = nominal angle of contact.
D = diameter of outer race.

2.3.4 Life of bearing in revolutions

The life of an individual ball bearing may be defined as the number of revolutions which the bearing runs before the first evidence of fatigue develops in the material of one of the rings or any of the rolling elements.

The rating life of a group of apparently identical ball bearing as the number of revolutions that 90% of a group of bearings will complete or exceed before the first evidence of fatigue develops:

$$L_n = \left(\frac{C_d}{P}\right) = \frac{14.8}{2} = 12.3 \tag{13}$$

P = weight of pulley + weight of shaft.
P = total load (.400 + .800) = 1.2 kg.
Life of bearing = 1800 million revolution.

2.3.5 Bearing number

Bore diameter = 20 mm.
Outer race diameter = 47 mm.
ISI no. = 20BC02.
Bearing of basic design no. = 04.
Basic capacity of dynamic load = 1000 kg.

2.3.6 Bearing socket

The outer diameter of the bearing race is 47 mm.
The diameter of the socket bore is 47 mm (**Figure 8**).

2.4 Design of driver and driven pulleys

The pulley was used to transmit power from one shaft to another by means of a flat belt, V-belt or rope.

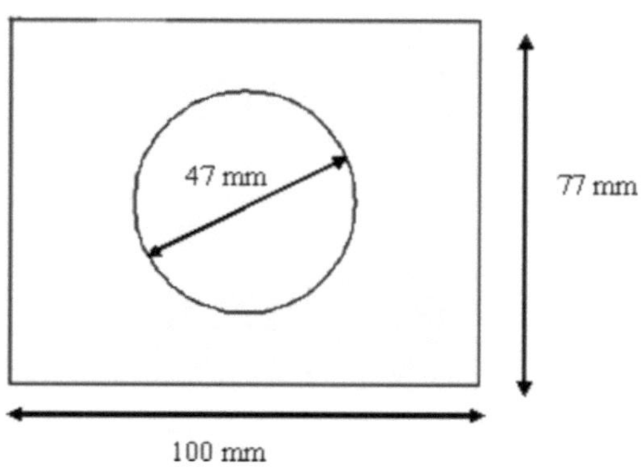

Figure 8.
Bearing socket.

2.4.1 Diameter of driver pulley

$$D = (110\text{–}113) \times \sqrt[3]{\frac{P(KW)}{2\pi N_{max}}} \quad (14)$$

$$D = 113 \times \sqrt[3]{\frac{.75}{2 \times \pi \times 1380}}.$$

D = 5.3 cm.
D = 6 cm (consider) = 60 mm (**Figure 9**).

2.4.2 Diameter of driven pulley

The velocity of driven pulley is increased by 1.4 times (**Figure 10**):

$$\text{velocity ratio} = \frac{N_2}{N_1} = 1.4 \quad (15)$$

$$\text{Diameter of driven pulley} = \frac{60}{1.4} = 43\ \text{mm}.$$

2.4.3 Thickness of web and depth of groove

$$T = 0.252\sqrt{D} + 1.6\ \text{mm (for light duty pulley)} \quad (16)$$

$T = 0.252\sqrt{60} + 1.6 = 3.55$
$T \approx 4$ mm.
Depth of groove on driver and driven pulley, where the rope is bounded =3 mm.
Angle of taper on groove = 5°.

2.4.4 Dimension of collar

A collar is used for locking of pulley on the shaft. The pulley is locked on the shaft by bolt. So, the pulley is not coming out when the shaft is rotating.

Diameter of shaft of driver pulley = 19.19 mm.

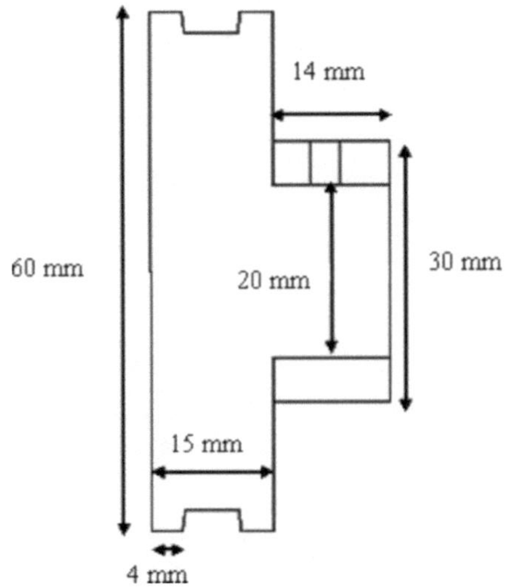

Figure 9.
Driver pulley.

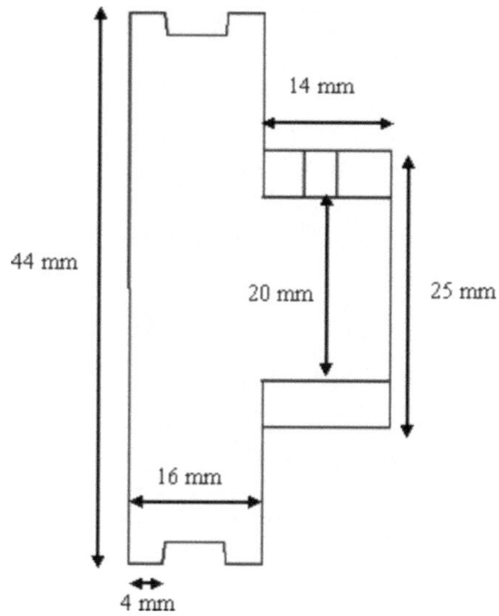

Figure 10.
Driven pulley.

Diameter of shaft of driven pulley = 20.00 mm.

Then,
Diameter of bore of collar of driver pulley = 20 mm.
Diameter of bore of collar of driven pulley = 20.02.
Length of collar of driver pulley = 14 mm.
Length of collar of driven pulley = 14 mm.
Diameter of tap = 4 mm (for both).

2.5 Design of milling socket

Material of socket = G.I.
Diameter of pipe = 27 mm.
Wall thickness of pipe = 3 mm.
Length of socket = 130 mm.
Morse taper = 1.5°.
Length of morse taper = 100 mm.
Material of inserted piece = M.S.
Length of inserted piece = 30 mm.
Diameter of through hole in inserted piece = 21 mm.
Internal threading = M10 × 1.5.
Length of thread = 15 mm (**Figure 11**).

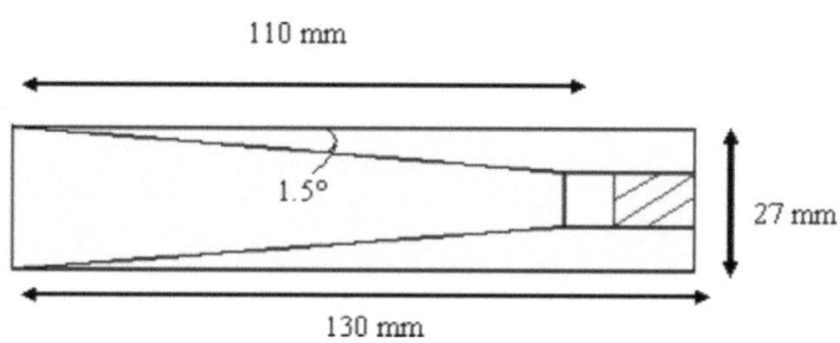

Figure 11.
Milling socket.

3. Experiment

The setup that was fabricated, with the use of this setup metallic foam, has been produced. For producing metallic foam, calcium carbonate is added into liquid aluminum and was stirred. Gases are formed at the time of stirring, and gas is entrapped into the liquid aluminum, and holes (porosity) are created into the aluminum. The foam that was formed can be cooled down, and after cooling the metallic foam can be taken out from the mold. Six experiments were done, and how much amount of aluminum and calcium carbonate can be taken is given below.

3.1 Experiment no. 1

Weight of aluminum = 887 g.
Amount of $CaCo_3$ = 50 g (approximate).

This was the first experiment and the amount of aluminum was 887 g. The amount of calcium carbonate added was 50 g (approximate), and after liquid aluminum was stirred and foam was developed. The foam was cooled down, and after cooling metallic foam was taken out from the mold. The porosity of foam was very low, so it was improved in the next experiments.

3.2 Experiment no. 2

Weight of aluminum = 900 g.
Amount of $CaCo_3$ = 30 g.

In first experiment, a satisfactory result was not obtained, so a second experiment will be performed. In this experiment the amount of aluminum was 900 g. The amount of calcium carbonate that was added was 30 g. It was added in three parts. The first part of calcium carbonate was in the mold before pouring aluminum. The second part of calcium carbonate was added when molten aluminum was poured into the mold. The remaining 10 g of aluminum was added at the time of stirring. The stirring time of a molten metal was 67 s. When calcium carbonate was added into molten aluminum, viscosity of aluminum was increased. After stirring, the stirrer was taken out from the mold and was cooled down. After cooling the foam was taken out from the mold. Metallic foam can be cut into pieces with the

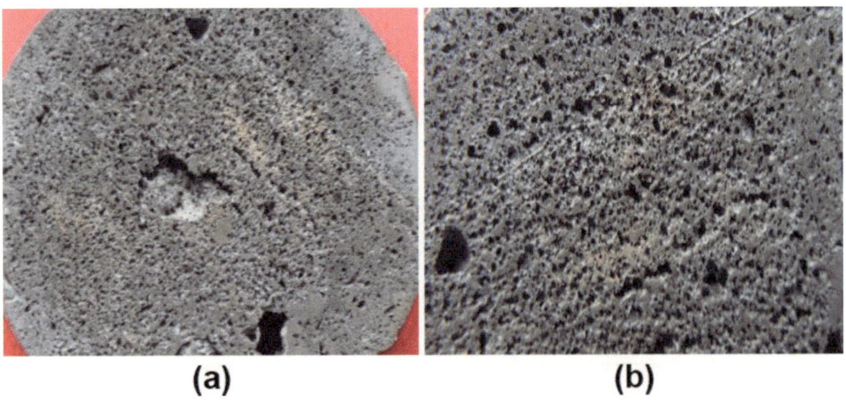

Figure 12.
Microstructure of metallic foam sample 2. (a) Microstructure of metallic foam; (b) section of metallic foam.

help of a wire cut electric discharge machining. The microstructure of metallic foam is given in **Figure 12a** and **b**.

3.3 Experiment no. 3

Weight of aluminum = 665 g.
Amount of $CaCo_3$ = 30 g.

Experiment 3 was not successful because calcium carbonate was directly added into the molten metal at the time of melting. All gases were escaped, and foam was not produced.

3.4 Experiment no. 4

Weight of aluminum = 969 g.
Amount of $CaCo_3$ = 20 g.

This was the fourth experiment. In the previous experiments, some ideas are obtained, and some standardization step was taken. The amount of aluminum was 969 g and the amount of calcium carbonate was 20 g (2% of weight). The process of producing metallic foam was shown in **Figures 13–19**. In **Figure 13** the setup is shown, which was used for producing metallic foam. In **Figure 14** aluminum metal was melt in a pit furnace. When aluminum melted, a slag was formed due to the impurities in the metal. In **Figure 15** mold was there in which molten metal was pouring. The mold was preheated with the help of a torch of gas welding. Preheating was required to reduce the temperature difference. In the mold there was one hole in the bottom. It was taken out molten metal after the foam was produced. In **Figure 16** the molten metal was poured into the mold. 20 g of calcium carbonate is divided into three parts. One part of 6 g was drawn before the molten metal is poured into the mold. When half of molten motel was poured, 7 g of calcium carbonate was added. In **Figure 17** the molten metal was stirred and the remaining amount of calcium carbonate was added. The stirring time was 60 s. After stirring, the stirrer was taken out. In **Figure 18** foam was generated due to the formation of gases, and these gases were entrapped in the aluminum. When the

Figure 13.
Setup for making a foam.

Figure 14.
Aluminum melted in a pit furnace.

Figure 15.
Mold.

Figure 16.
Aluminum poured into the mold.

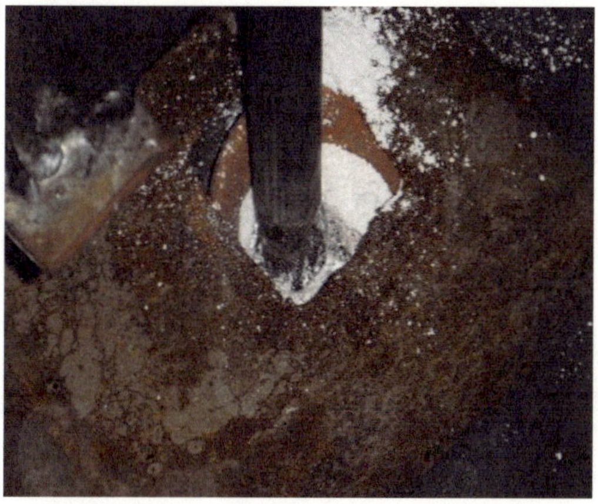

Figure 17.
Add calcium carbonate and stir it.

Figure 18.
Metallic foam.

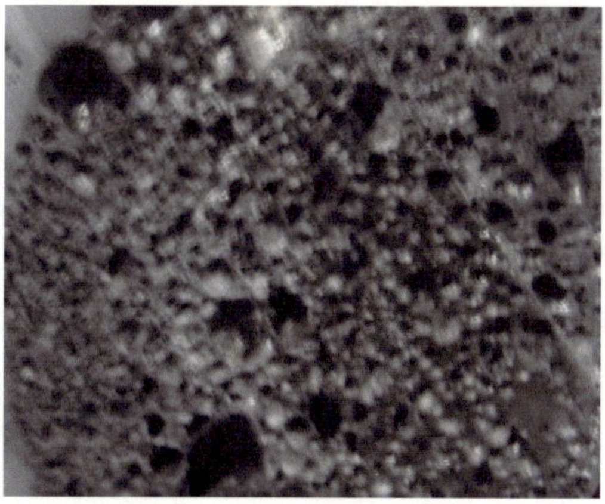

Figure 19.
Microstructure of metallic foam sample 4.

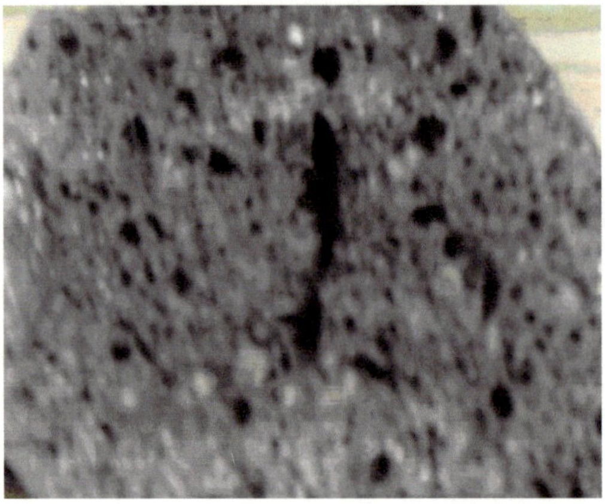

Figure 20.
Microstructure of metallic foam sample 4.

aluminum was cooled down, casting can be taken out from the mold. In **Figure 19** the microstructure of metallic foam that was produced was shown.

3.5 Experiment no. 5

Weight of aluminum = 776 g.
Amount of $CaCo_3$ =?

Experiment 5 was not completed because the molten metal was drained out from the hole.

3.6 Experiment no. 6

Weight of aluminum = 330 g.
Amount of $CaCo_3$ = 6 g.

This was the sixth experiment in which the amount of aluminum was 330 g and amount of calcium was 6 g (2% of weight). The stirring time was 60 s. All procedure was the same as described above. The drainage time of the molten metal from the bottom hole was 60 s. The microstructure was shown in **Figure 20.**

4. Result

Test the aluminum that was used is pure or not, Archimedes' principle was used. In Archimedes' principle water taken in a bucket and that bucket was full of water at the top level. Then a piece of aluminum that was used for producing metallic foam was dipped into the water, and the water that was coming out from the bucket was collected into the next bucket. The weight of water was calculated and calculations that are given below are performed:

Weigh of aluminum piece = 85 g.
Weight of water displaced out = 35 g.

$$\text{Density of water } \rho_w = \frac{m_w}{v_w} \tag{17}$$

$v_w = \frac{m_w}{\rho_w} = 3.5 \times 10^{-5}\ \text{m}^3$.

The volume of water was equal to the volume of the aluminum piece:

Density of aluminum: $\rho_{\text{Al}} = \frac{m_{\text{Al}}}{v_{\text{Al}}}$.

$\rho_{\text{Al}} = 2.42 \times 10^3\ \text{kg/m}^3$.

The density of pure aluminum was $2.319 \times 10^3\ \text{kg/m}^3$.

So, there were some impurities in the aluminum that are used in this research.

5. Density of metallic foam

Cut a rectangular piece of metallic foam and take the weight of that piece. Calculate the dimensions of piece also and find the volume of that piece:

Dimension of piece = 65 × 20 × 15 mm.

Volume of piece = $1.9500 \times 10^{-5}\ \text{m}^3$.

Weight of metallic foam piece = 8.25 g = 0.00825 kg.

$\rho_f = 0.4230 \times 10^3\ \text{kg/m}^3$.

6. Conclusion

There is a large potential for metal foam applications. However, there are several deficiencies with the current technology. With these opportunities for the improvement identified, research objectives were set to design a new foam product and develop a new method of production to overcome the present shortcomings. Given the current results of this research work, it is evident that these areas for improvement have been addressed, and the knowledge gained from this work lends valuable information to plan future actions for further improvement to this product.

The system was designed, fabricated and controlled for aluminum foam generation. The obtained foam was tested, and the density of metallic foam was $0.4230 \times 10^3\ \text{kg/m}^3$.

Author details

Shyam Sunder Sharma
Department of Mechanical Engineering, Manipal University Jaipur, India

*Address all correspondence to: shyamsunder.sharma@jaipur.manipal.edu

References

[1] Jung A. Open-cell aluminium foams with graded coatings as passively controllable energy absorbers. Materials and Design. 2015;**87**:36-41

[2] Sun Y, Burgueno R, Wang W, Lee I. Effect of annealing on the mechanical properties of nanocopper reinforced open-cell aluminum foams. Material Science Engineering. 2014

[3] Banhart J. Metal foams: Production and stability. Advanced Engineering Materials. 2006;**8**(9):781-794. Berlin

[4] Banhart J, Baumeister J. Production methods for metallic foams. In: Materials Research Symposium Proceedings. Vol. 521. 1998

[5] Matijasevic B, Banhart J. Improvement of Aluminium Foam Technology by Tailoring of Blowing Agent. Berlin, Germany: Elsevier; 2005

[6] Ashby MF, Evans A, Fleck NA, Gibson LJ, Hutchinson JW, Wadley HNG. Metal Foams: A Design Guide. Massachusetts: Butterworth-Heinemann; 2000

[7] Gibson LJ, Ashby MF. Cellular Solids, Structure and Properties. 2nd ed. Cambridge: Cambridge University Press; 1997

[8] Bhatt A, Khanna M, Pimoli BS. Metal foaming of aluminium alloys. IOSR Journal of Mechanical and Civil Engineering. 2015;**12**(1):40-44. Ver. IV.e-ISSN: 2278–1684, p-ISSN: 2320-334X

[9] Sharma SS, Rajpoot YS. Development of aluminum metal foam using blowing agent. In: IOP Conference Series: Materials Science and Engineering; Vol. 377(1). IOP Publishing; 2018. p. 012150

[10] Banhart J. Manufacture, characterisation and application of cellular metals and metal foams. Progress in Materials Science. 2001; **46**(6):559-632

Chapter 6

The Improvement of the Quality of Construction Foam and Non-Autoclave Foam Concrete on Its Basis through the Introduction of Nanosize Additives

Anastasia Sychova, Larisa Svatovskaya and Maxim Sychov

Abstract

In order to improve the quality of a construction foam on a protein basis for non-autoclaved foam concrete, a proposal has been made to increase its stability by introducing nanosize additives—SiO_2 and $Fe(OH)_3$ sols. It is shown that the effect obtained is connected with various stabilization mechanisms. It is stated that these mechanisms are connected with different energies of chemical bonds formed between the molecules of the foaming agent and the injected sols. By means of electron microscopy, it is stated that the growth of foam stability is connected with an increase in the foam film thickness by one order. An increase in the coefficient of the foam resistance in the cement paste is shown. The stabilization of the construction foam leads to the possibility of using foam concrete hardening accelerators without destroying its structure. The resulting foam concrete is proved to get the increased compressive and bending tensile strength and reduced thermal conductivity and shrinkage in drying. The porosity of the foam concrete obtained is tested by means of mercury porometry. Its phase composition is investigated by X-ray phase and derivatographic analysis.

Keywords: construction foam, foam stability, protein foaming agent, nanosize additives, sol SiO_2 and $Fe(OH)_3$, non-autoclaved foam concrete

1. Introduction

Under the stability of the foam, one understands its ability to maintain the dispersed composition, volume, and ability to prevent delamination. Foam stability is directly connected with the properties of liquid thin layers (foam films) which determine the structure of the foam. To evaluate quantitatively the stability of the foam, the rate of its destruction is determined [1].

Foaming agents used for the production of foam concrete of different hardening are considered in [2, 3]. The paper presents the classification of foaming agents according to their chemical characteristics [4, 5]. Properties of various foams, the bases of their obtaining, and destruction are investigated in the works [6, 7]. For foam concrete on the cement binder, the influence of the foaming agent used on the degree of cement hydration is considered. It is shown that the protein-based foaming agent is the best one [8].

Instability and destruction of the construction foam is one of the causes of volume instability in foam concrete. Especially sharply this question is on the lightweight foam concretes, because the volume of the foam in their composition is up to 90% [9].

In this regard, one of the ways to improve the stability of the foam concrete mixture can be the stabilization of the construction foam introduced into the cement-sand paste.

At present there are various ways to improve the stability of foams based on various stabilization mechanisms. The use of additives (glycerin, methylcellulose, ethylene glycol) increases the viscosity of foaming agent solutions and slows down the liquid slug from foam films [10]. Also, for the stabilization of foams, it is possible to use substances that contribute to the formation of colloidal particles in films, preventing their dehydration. This group includes gelatin, joiner's glue, starch, and polysaccharides [10, 11].

For the production of thermal insulation materials, it is also recommended to use substances polymerized in the foam as stabilizers. Such additions strengthen the foam film significantly. They include polymer compositions based on synthetic resins and latex [12].

In addition, there are some ways of improving the stability of the foam based on other mechanisms of stabilization.

In [13] the influence of spherical monodisperse SiO_2 particles with a diameter of 20 to 700 nm on the stabilization of foam based on sodium sulfonate was investigated. The positive effect obtained in this case is shown.

The question of obtaining stable foams formed from water dispersions of laponite modified by hexylamine is considered in [14]. It is shown that such a composition "surrounds" foam bubbles with a thin layer, providing a stabilizing effect.

It is also known that the foam can be stabilized by hydrophobic polymer particles with a diameter of less than 1 μm and a length of several tens of μm [15]. The stabilization effect is connected with the formation of dense thick layers of these particles around the foam films.

The paper [16] provides information that submicrometer-sized polystyrene particles in combination with poly[2–(diethylamino)ethyl methacrylate] of various degrees of polymerization—30, 60, and 90—can be used as stabilizers for foam. A higher degree of polymerization results in the highly stable foams.

It is also known that the use of hydrophilic silica particles and liquid paraffin in the foam increases its stability considerably [17]. It is shown that the stabilizing effect is connected with the adsorption of silica particles and oil droplets on the air-water layer of the foam film.

The publication [18] provides information that a significant increase in the stability of the foam can be achieved by the introduction of silica particles of the micrometer size. It is also proposed to use spherical silica particles with a diameter from 150 to 190 nm with their subsequent modification by silane substances [19]. It is established that the hydrophobicity of the surface achieved is a key factor affecting the stability of the foam.

In [20] the data confirming the fact that the stabilization of foams by solid particles is possible and results in a good effect are given. The result depends on the packaging of the particles on the surface of the foam films. The denser the packaging, the better the effect. Also, on the basis of calculations, it is shown that for stabilization of foams on the water basis and on the basis of the liquid, aluminum particles with a diameter less than 3 and 30 μm, respectively, are required.

In [21] the method of determining the coefficient of effectiveness and critical coverage as well as the coefficient of adsorption of particles on the interface of foam bubbles is considered. The possibility of obtaining the most stable foam is stated to depend on them.

The possibility of obtaining porous ceramics from a foamed suspension based on Al_2O_3–TiO_2/ZrO_2–SiO_2 is discussed in [22]. Suspension particles have a stabilizing effect and allow to obtain foam with air content up to 87%.

The carried out analysis suggests that the increase in the stability of the foam is mainly due to the mechanism of "armoring" the surface of the foam bubble by the injected solid particles. This process prevents the liquid from leaking out of the foam film under the influence of gravity and so prevents its subsequent destruction. The mechanisms of foam chemical stabilization with the purpose of getting the foam concrete are not considered yet.

The use of additives based on SiO_2 nanoparticles in modern construction is well known [23–25].

One of the highly effective nano-additives used today is known to be silica (SiO_2) sol [26–28]. Sols are colloidal aqueous solutions containing nanosize particles (1–100 nm) [29]. Chemistry of colloidal silica and its applications are discussed in detail in literature and primarily in [30, 31].

As the literature review shows, there is no knowledge concerning the methods of stabilization of construction foam on a protein basis for foam concrete at the expense of the introduction of sols of different nature into its composition. Also, the problems of improving the physical, mechanical, and thermal properties of the foam concrete of non-autoclaved hardening obtained on the basis of such a foam are not considered.

2. Method

The main idea of the work is: it is possible to use SiO_2 and $Fe(OH)_3$ sols as stabilizers for protein foaming agent (PFA), as it is possible to form various chemical bonds between them, for example, hydrogenous, in the case of using SiO_2 sol [30], or covalent, in the case of using $Fe(OH)_3$ sol [12]. These bonds should contribute to the formation of strong spatial silicon- and iron-protein complexes which will increase the thickness of the foam film and prevent its destruction. This effect should preserve the volume of foam when it is introduced into the cement-sand mixture, as well as improve the physical, mechanical, and thermal characteristics of foam concrete and its products.

The expected chemical bonds formed in the system "aqueous solution of protein foaming agent, sol of different nature," are shown in **Table 1**. The spatial stabilizing complexes based on them are also shown. The formation of these complexes is assumed to provide the foam stability increase. The expected influence of these effects on the physical and technical properties and quality of non-autoclaved foam concrete and its products is described.

The foam stabilization effect can be important for the production of heat-insulating foam concrete of low density (class of average density $\leq D_{200}$), as it will prevent volume instability of raw foam concrete mixture. Also, this effect will make it possible to use hardening accelerators—electrolytes—in the composition of foam concrete which usually destroy it. In high-rise construction when the foam concrete mixture is fed vertically to a great height, foam stabilization will prevent its destruction under the influence of its own pressure in the falling pipeline [12].

It is known that when getting foam concrete products by cutting technology, there appear problems with cutting the mass of foam concrete—there arise chips

System "aqueous solution of protein foaming agent—sol of different nature"			
Stabilizer	**Stabilizing complex**	**Type of chemical bond**	**Expected effect**
SiO_2 sol	Silicon-protein complex 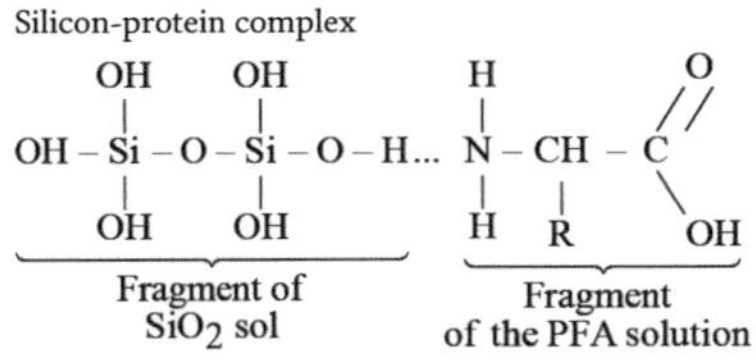	Hydrogenous H—N	• The foam stability increase • The increase of the foam stability rate in the cement paste and preservation of the concrete mix volume • The possibility of additives-electrolyte application • The increase of compressive and tensile strength of foam concrete • The reduction of thermal conductivity coefficient and shrinkage of foam concrete in drying • The quality category increase of the foam concrete products
$Fe(OH)_3$ sol	Iron-protein complex 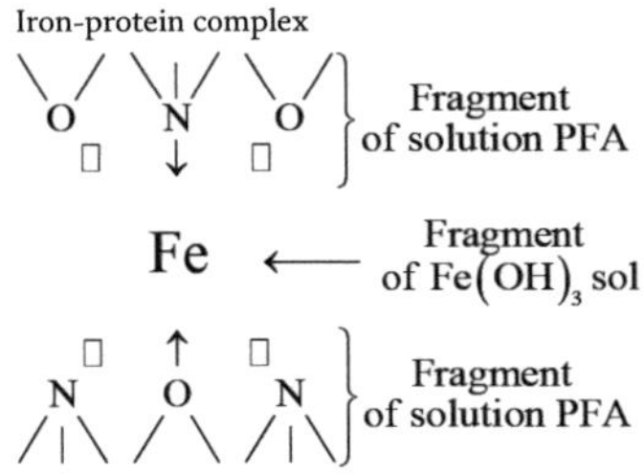	Covalent Fe — O, Fe — N	

Table 1.
The expected influence of sols of different nature on the quality of foam, foam concrete, and its products.

and other geometric defects. They reduce the quality category of foam concrete products to category II (according to GOST 31360-2007).

To obtain the first quality category is important because in carrying out a brickwork, it allows to place the foam concrete blocks on the construction glue (coefficient of thermal conductivity, $\lambda \approx 0.3$ W/(m•°C)), and not on the cement mortar $\lambda \approx 0.3$ W/(m•°C)). The use of construction glue is proved to increase the thermal insulating properties of masonry walls significantly. It is assumed that the stabilized foam and the use of the hardening accelerator will considerably increase the number of products of the first quality category.

3. Study of the stabilized foam properties

To confirm the stabilizing effect of SiO_2 and $Fe(OH)_3$ sols, the stability of the construction foam was investigated depending on the concentration of the dispersed phase of the sols in the solution of the protein foaming agent.

In the study, a protein foaming agent "Foamcem" was used as a foaming agent, on the basis of which a 3% aqueous solution was prepared. In addition, SiO_2 sol of the industrial production "SITEC" was used, its characteristics being shown in **Table 2**. Also, $Fe(OH)_3$ sol obtained in the laboratory was used.

$Fe(OH)_3$ sol was obtained by the following method: 5 ml of a 10% solution of iron chloride $FeCl_3$ was slowly poured into boiling water with a volume of 100 ml.

Title	Density (kg/m^3)	Concentration, ω=SiO_2 (%)	pH	Specific surface area, S (m^2/g)
KZ-1 "SITEC"	1165	25.6	3.2	120–140

Table 2.
Main characteristics of the industrial SiO_2 sol.

The stability of the foam was evaluated as the time of extraction (in minutes) of a half of the liquid phase from which the foam was prepared.

Foam stability in the cement paste was evaluated by the coefficient of the foam resistance. Determination of the resistance coefficient was made by mixing equal volumes of cement paste and foam for 1 min followed by measuring the volume of the porous paste. The resistance coefficient of foam in the cement paste is calculated as the ratio of the volume of the porous paste to the sum of the volumes of the cement paste and foam (with water/cement ratio = 0.4).

The results of the studies are shown in **Figures 1–4**. From the figures it can be seen that the stability of the foam stabilized with the SiO_2 and $Fe(OH)_3$ sols increases up to four times and the foam resistance coefficient in the cement

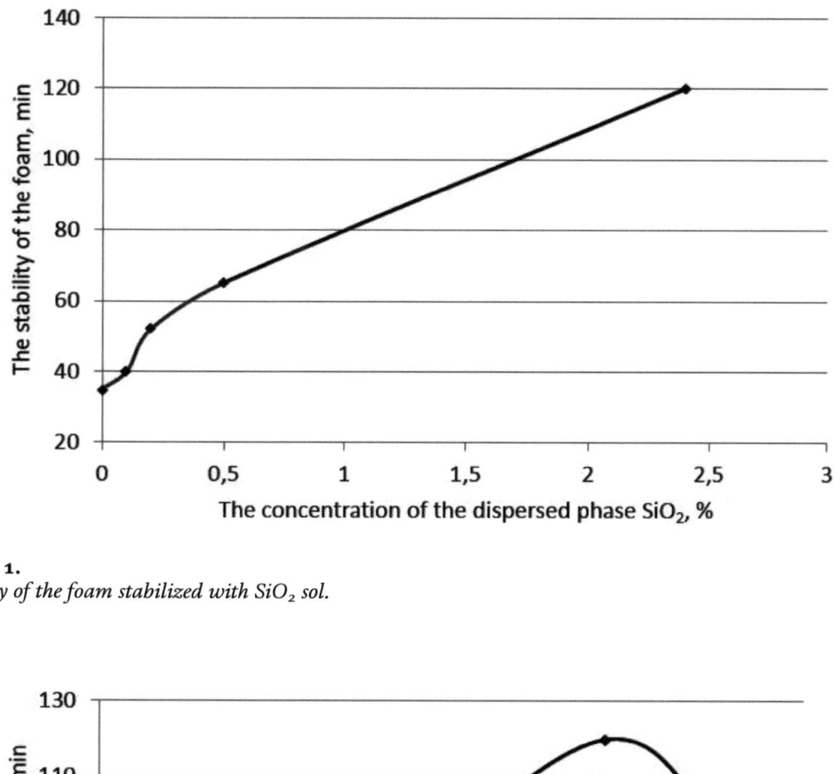

Figure 1.
Stability of the foam stabilized with SiO_2 sol.

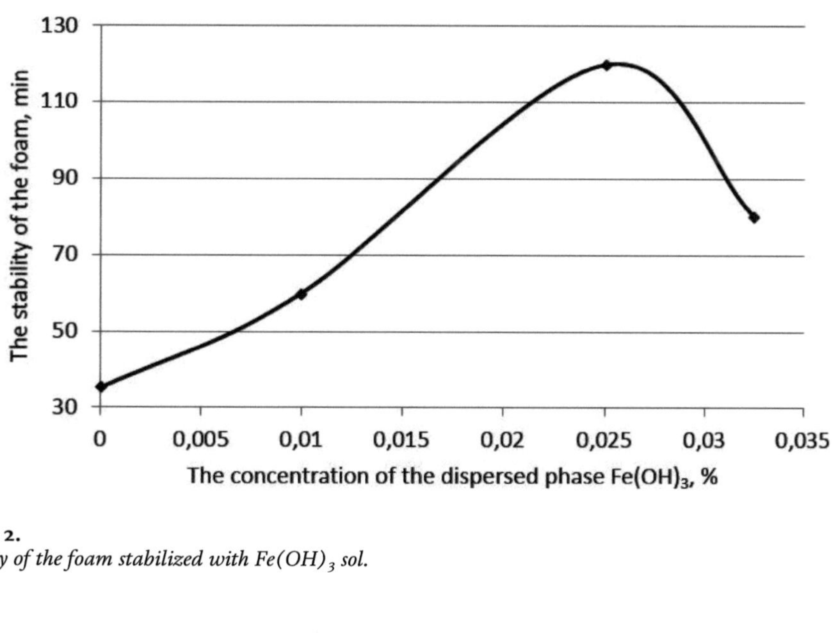

Figure 2.
Stability of the foam stabilized with $Fe(OH)_3$ sol.

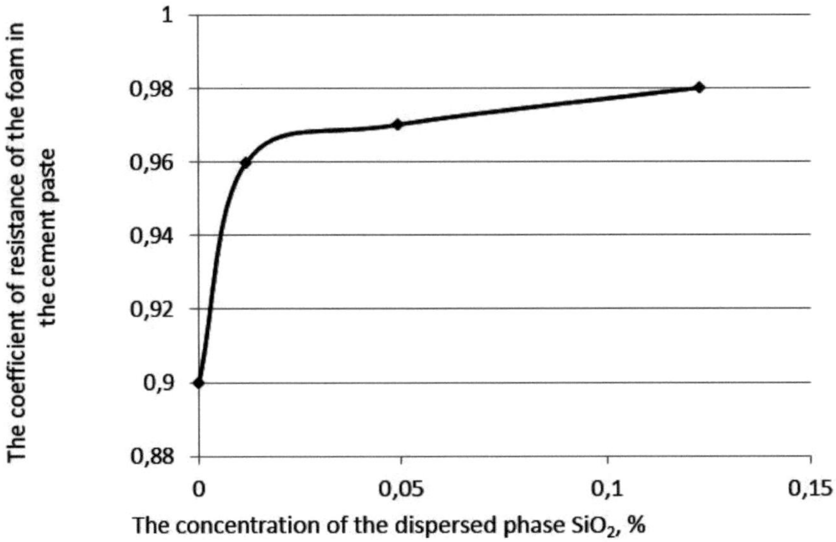

Figure 3.
The coefficient of resistance of the foam stabilized with SiO_2 sol in the cement paste.

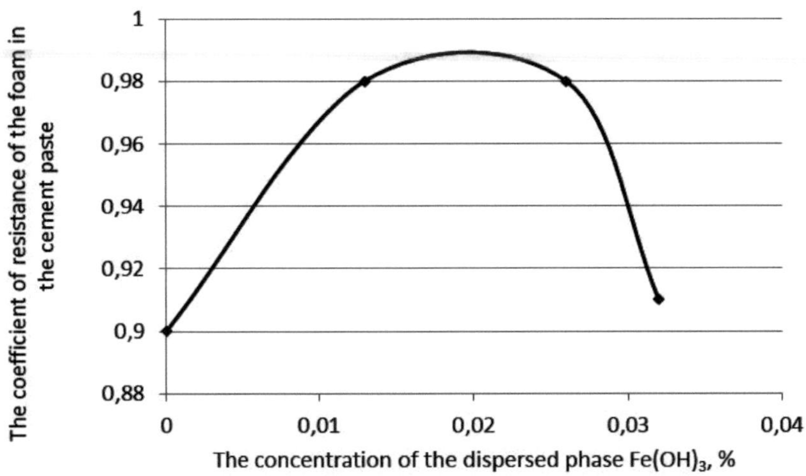

Figure 4.
The coefficient of resistance of the foam stabilized with $Fe(OH)_3$ sol in the cement paste.

paste increases from 0.9 up to 0.98. These results agree with the forecast as shown in **Table 1**.

Further, in order to clarify the stabilizing mechanism, the surface tension of the foaming agent solution was measured at different concentrations of the dispersed phase of SiO_2 sol, as shown in **Table 3** [10]. The table shows that with increasing the sol concentration, the surface tension practically does not change.

Further, the foam multiplicity (frequency rate) when introducing of SiO_2 and Fe $(OH)_3$ sols into the solution of the foaming agent was investigated. The foam multiplicity was determined by the ratio of the foam volume (l) to the foam solution volume (l). It was found that the multiplicity of the foam on the sol basis does not change, which correlates with the results of measuring the surface tension of the modified foam solutions. The value of the foam multiplicity was 13.

Further, in order to determine the possible chemical bonds formed between the molecules of the protein foaming agent and the injected sols, infrared Fourier spectroscopy of the following model systems was carried out: "aqueous solution of the protein foaming agent" and "aqueous solution of the protein foaming agent—the injected sol" (**Figures 5** and **6**).

From spectrum No. 2, **Figure 5**, the shift and broadening of the band in the region of 3311 cm^{-1} corresponding to the valence vibrations of the hydroxide (OH^-) groups in comparison with spectrum No. 1 are seen. This effect can characterize the occurring hydrogen bonds between the nitrogen atom in the protein and hydrogen of the OH^- group of SiO_2 sol with the formation of a silicon-protein complex, as shown in **Figure 7a**.

In both spectra (**Figure 6**) the bands in the region of adsorption 1630–1510 cm^{-1} correspond to the deformation vibrations of the carbonyl group C=O of protein. The shift and broadening of this line can be observed on spectrum No. 1, which may correspond to the formation of a covalent bond of iron ion (III) with oxygen of the amide group in the composition of the iron-protein complex formed, as shown in **Figure 7b**. The region of 1150 cm^{-1} corresponds to the deformation vibrations of the NH—C=O group, the shift and broadening of this band on spectrum No. 1 indicate a possible covalent bond of the iron ion (III) with nitrogen in the composition of the iron-protein complex formed, as shown in **Figure 7b**.

Thus, it can be concluded that the decoding of IR spectra of foaming agent solutions stabilized with various sols confirms the assumption expressed in **Table 1**

Concentration of the dispersed phase of the sol in the foaming agent solution (%)	0	0.306	0.610	3013
The surface tension coefficient of the foaming agent solution, σ^*, 10^{-3} (J/m^2)	52.9 ± 0.8	52.9 ± 0.8	53.2 ± 1.2	52.1 ± 1.1

**The value of the surface tension coefficient at the ambient temperature of 21°C.*

Table 3.
Values of the surface tension coefficient of foaming agent solution stabilized with SiO_2 sol.

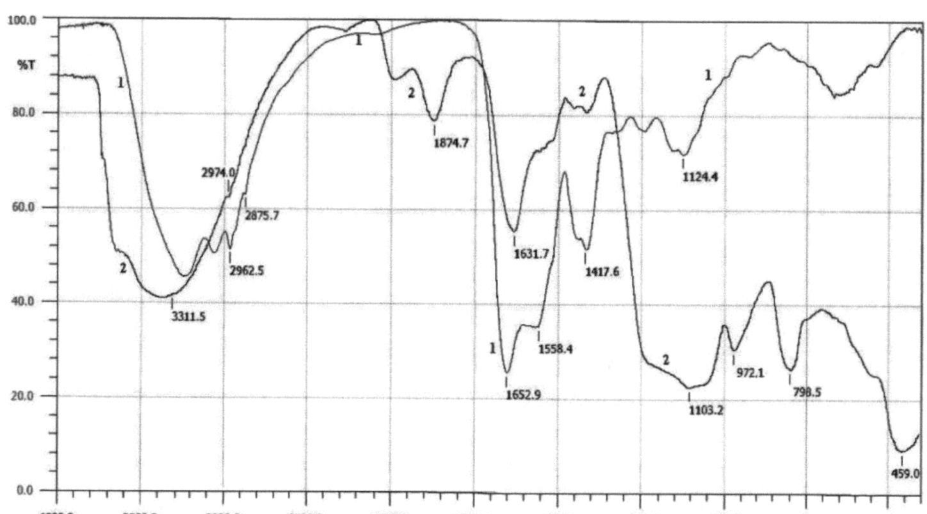

Figure 5.
IR spectra: No. 1, the system "aqueous solution of the protein foaming agent." No. 2, the system "aqueous solution of the protein foaming agent—SiO_2 sol."

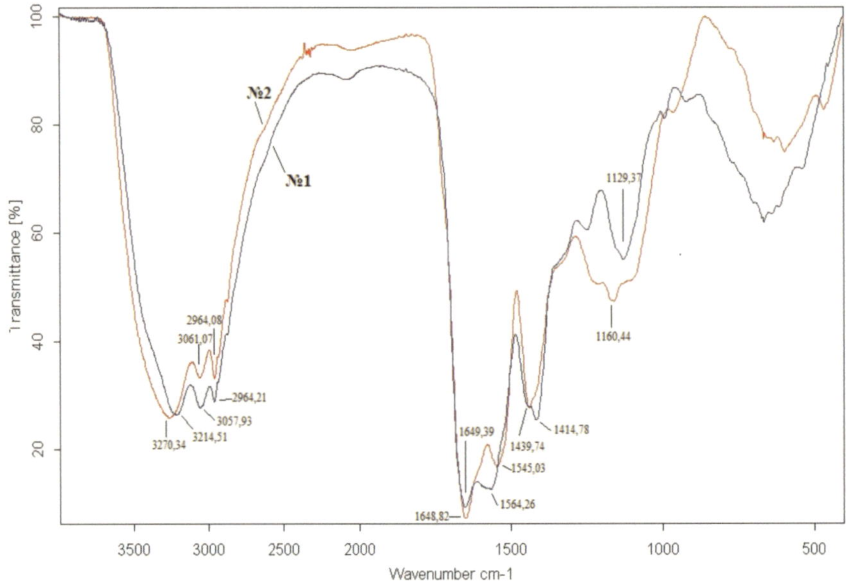

Figure 6.
IR spectra: No. 1, the system "aqueous solution of the protein foaming agent—Fe(OH)$_3$ sol." No. 2, the system "aqueous solution of the protein foaming agent."

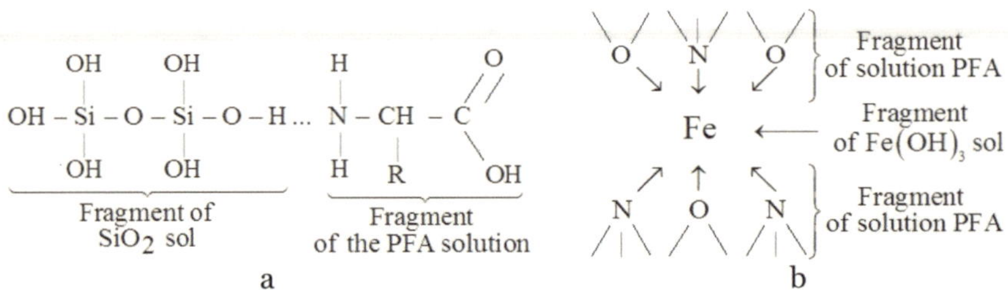

Figure 7.
Spatial stabilizing complexes formed by the introduction of various sols into the solution of protein foaming agent: (a) silicon-protein complex and (b) iron-protein complex.

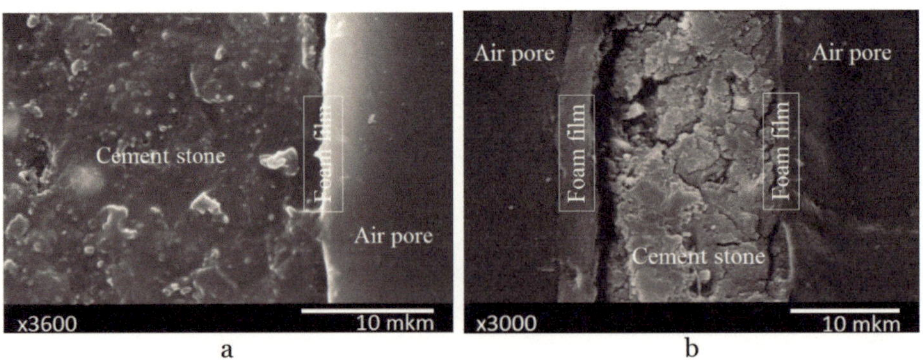

Figure 8.
Electron microscopy of the foam concrete samples of the medium density D600: (a) control sample and (b) sample based on foam stabilized with SiO$_2$ sol.

concerning the formation of hydrogenous and covalent chemical bonds in the composition of the stabilizing silicon- and iron-protein complexes.

Further, it was assumed that the appearance of chemical bonds and spatial stabilizing complexes should increase the thickness and strength of the foam film, which results in the increase of the foam stability. For this purpose, electron microscopy of samples of the foam concrete with medium density D600 was carried out: a control sample and a sample prepared on the basis of the foam stabilized with the SiO_2 sol (**Figure 8**).

From the pictures it can be seen that the thickness of the foam film in the control sample is ~450 nm and in the sample based on the stabilized foam is ~3.5 μm, i.e., increase by one order. This result can explain the stabilizing effect and the increase in foam stability as well as confirm the assumptions made.

4. Research of the foam concrete properties of various average densities on the basis of the stabilized foam

Further the possibility of receiving heat-insulating non-autoclaved foam concrete of average density D_{200} on the basis of the foam stabilized with SiO_2 sol was investigated.

For light thermal insulating non-autoclaved foam concrete, one of the significant problems is the reduction of the volume of the foam concrete mixture because of the mixture destruction, which results in the deviation from the projected average density and uneven structure of the material and exerts a negative influence on the properties of thermal insulating foam concrete.

It was assumed that a more stable foam will allow to avoid the destruction of the mixture, to provide the necessary density of the foam concrete, and to obtain a reduced thermal conductivity coefficient.

The composition of the foam concrete with medium density D_{200} is shown in **Table 4**. Portland cement CEM 42.5 was used as a binder, dolomitized limestone was used as a filler, protein foaming agent "Foamcem" was used as a foaming agent, and SiO_2 sol of industrial production was used as a stabilizer (SITEC company).

To assess the stability of the foam concrete mixture, the volume instability of the foam concrete (mm) was measured at different contents of the dispersed phase of the sol in the foam. Volume instability was measured after 24 hours of the foam concrete hardening (**Figure 9**). From the figure it is seen that the use of the stabilized foam reduces the volume instability up to 0 when the concentration of the dispersed phase of sol in the foam is at least 0.2%. The coefficient of thermal conductivity of foam concrete of average density D_{200} at the design age was $\lambda = 0.04$ W/(m•°C); for comparison, $\lambda_{air} = 0.029$ W/(m•°C).

Further the physical-technical and thermal insulating properties of foam concrete and its products after the introduction of the stabilized foam into its composition were evaluated.

The compositions of the foam concrete mixtures of different average densities are shown in **Table 5**. During the preparation of foam concrete mixtures, 3% aqueous solution of foam on a protein basis stabilized with different sols was used. The samples were being solidified under normal conditions for 28 days.

During the experiment it was expected that a more stable foam will allow to use the electrolyte additives to activate the hardening of cement and to obtain the improved physical and mechanical characteristics of foam concrete and its products. In the case of using a conventional foam solution, such additives destroy the

Cement (kg)	Filler (kg)	Water (l)	Foaming agent (l)	Stabilizer (SiO_2 sol) (kg)
125	45	102	2.52	0.3

Table 4.
Consumption of the materials for 1 m^3 of foam concrete of average density D_{200}.

foam. Sodium chloride (NaCl) in an amount of 5% by weight of cement was used as a hardening activator. The obtained characteristics of foam concrete are shown in **Table 6**.

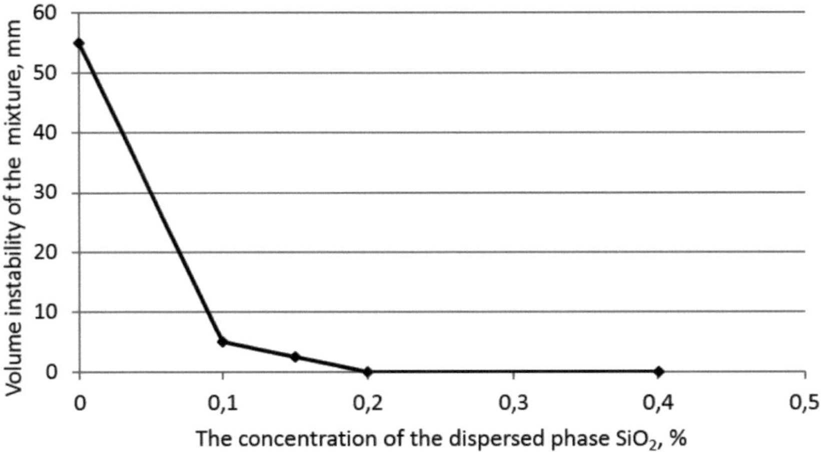

Figure 9.
Dependence of the volume instability of the foam concrete mixture on the content of the dispersed phase of SiO_2 *sol in the foaming agent solution.*

Class by average density	Cement (kg)	Filler (kg)	Water (l)	Foaming agent (l)	Stabilizer (l)	
					SiO_2 sol	$Fe(OH)_3$ sol
D_{400}	330	50	152	2.1	1.5	6.1
D_{500}	370	100	183	1.98	1.44	5.7
D_{600}	400	170	211	1.75	1.3	5.0

Table 5.
Consumption of materials per 1 m³ of foam concrete of different average densities.

Class by average density	Foam type	Stabilizer	Compressive strength, MPa/%	Bending tensile strength, MPa/%	Thermal conductivity, λ, W/(m•°C)/%
D_{400}	Control	—	0.8/100	0.45/100	0.100/100
	Stabilized	SiO_2 sol	1.2/150	0.76/169	0.086/86
		$Fe(OH)_3$ sol	1.2/150	0.66/147	0.09/90
D_{500}	Control	—	1.3/100	0.69/100	0.120/100
	Stabilized	SiO_2 sol	1.9/146	1.14/165	0.101/84
		$Fe(OH)_3$ sol	1.7/131	0.88/128	0.107/89
D_{600}	Control	—	1.7/100	0.88/100	0.140/100
	Stabilized	SiO_2 sol	2.3/135	1.38/158	0.117/84
		$Fe(OH)_3$ sol	2.2/129	1.15/131	0.132/94

Table 6.
Physical and mechanical characteristics of foam concrete samples of average density D_{400}*–*D_{600} *prepared on the basis of the stabilized foam.*

Stabilizer	Class of frost resistance
—	F15
Sol SiO_2	F35
Sol $Fe(OH)_3$	F35
All studies were conducted in accordance with GOST 25485-89.	

Table 7.
Frost resistance samples of the foam concrete of average density D600 prepared on the basis of the stabilized foam.

When assessing the physical and mechanical characteristics of the foam concrete samples obtained on the basis of a stabilized foam, the following results were got:

- The use of electrolytes does not destroy the stable foam.
- The compressive strength of the samples with medium density D_{400}–D_{600} at the age of 28 days of normal hardening increases up to 50% compared to the control sample, and the tensile strength at bending increases up to 69%.
- The thermal conductivity coefficient of the samples with additives decreases up to 16% in comparison with the control sample.

Also, for samples of foam concrete of average density D_{600}, a study of their frost resistance was carried out. It showed an increase in the class of frost resistance from F15 to F35, as shown in **Table 7**.

5. Physicochemical studies of the composition of the obtained foam concrete and its porous structure

At the next stage of the work, physical and chemical studies of samples of foam concrete with medium density D_{500}, prepared on the basis of stabilized foam and NaCl additives, were carried out: X-ray phase and differential thermal analysis. Three samples of foam concrete were studied: No. 1, control; No. 2, based on the foam stabilized with SiO_2 sol and with the addition of NaCl; and No. 3, based on the foam stabilized with $Fe(OH)_3$ sol and with the addition of NaCl.

In all samples X-ray phase analysis showed the presence of reflexes corresponding to β-SiO_2 with d/n (interplanar spacing) = (3.337, 2.447, 2.280, 2.119, 1.657; 1.539) Å, as well as reflexes corresponding to $Ca(OH)_2$, d/n = (3.114, 2.625, 1.926, 1.675) Å, low-basic hydrosilicate C_6S_6H d/n = (3.030, 2.033, 1.95) Å, and hydrosilicate C_2SH_2 (d/n = (3.030, 2.765, 1.830, 1.565) Å). In the X-ray spectra of samples No. 2 and No. 3, new lines belonging to the low-basic hydrosilicate $C_3S_2H_3$ (d/n = 2.88; 2.766; 2.152; 1.973; 1.793; 1.627 Å) appear. The radiographs of samples No. 2 and No. 3 show the lines characterizing the dolomitized limestone. Alite analytical line (C_3S, d/n = 1.76 Å) is present only in the control sample; in other samples, it does not manifest itself, which indicates a deeper degree of cement hydration in them. The formation of additional low-basic hydrosilicates with increased strength, as well as the absence of an analytical line of alite on the radiographs, can explain the increase in the strength of foam concrete samples No. 2 and No. 3.

The derivatographic analysis, **Table 8**, confirmed the data of X-ray phase analysis and showed that the total mass loss of samples based on the stabilized foam and

NaCl additives increases by 20%; besides that, a new phase of low-basic hydrosilicate $C_3S_2H_3$ (endothermic effect in the region of 350–400°C) appears in the samples, which also confirms the increase in the strength of the foam concrete samples.

Then, in order to assess the porous structure of foam concrete based on the foam stabilized with SiO_2 sol, porosity of the samples was investigated by means of mercury porometry (**Figure 10**).

The figure shows that the specific surface area of the pores in the sample based on the stabilized foam is twice that of the control sample. This may be due to the fact that such a foam is more stable in preparing a foam concrete mixture and in subsequent hardening the fine porous structure of the material is retained. The studies of the macroporous structure of foam concrete confirm this conclusion, as shown in **Figure 11**.

No.	Stabilizer	Endothermic effects, °C					Total mass loss on the effects, mg	The total mass loss by the sample, mg
		Mass loss, mg						
	Additive	(130–170)	(350–400)	(520–580)	(750–880)	(930–960)		
1.	—	88	—	24	14	—	126	179
2.	$Fe(OH)_3$ sol / NaCl	86	16	20	33	28	183	213
3.	SiO_2 sol / NaCl	90	18	19	40	30	197	227

Table 8.
Derivatographic analysis of foam concrete samples of average density D_{500}.

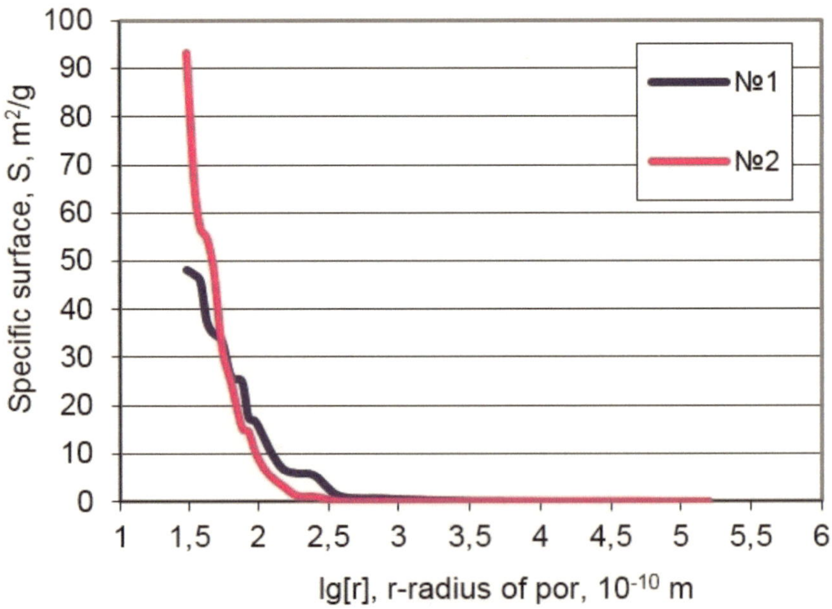

Figure 10.
The total specific surface of pores of foam concrete with average density of D_{500}: (1) a control sample, (2) a sample based on the foam stabilized with SiO_2 sol.

Figure 11.
Macroporous structure of samples: (a) a control sample and (b) a sample based on the stabilized foam.

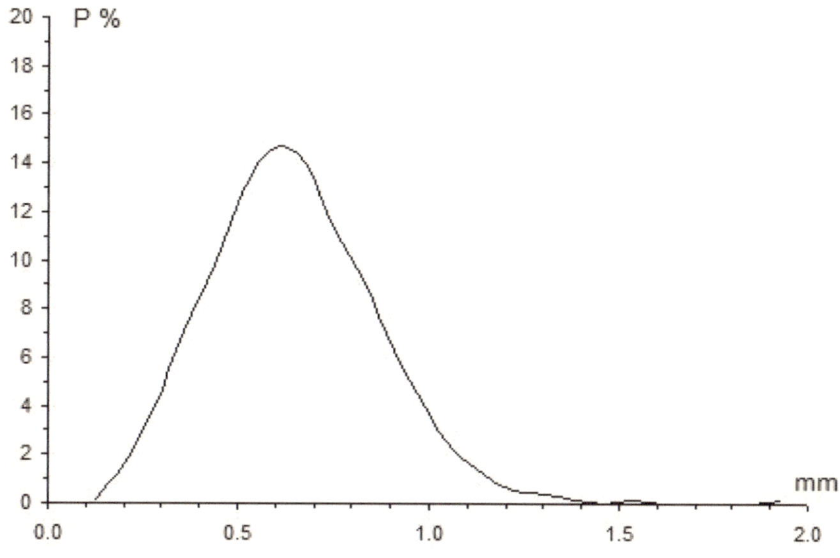

Figure 12.
Distribution of macropores according to their size for the control sample of foam concrete.

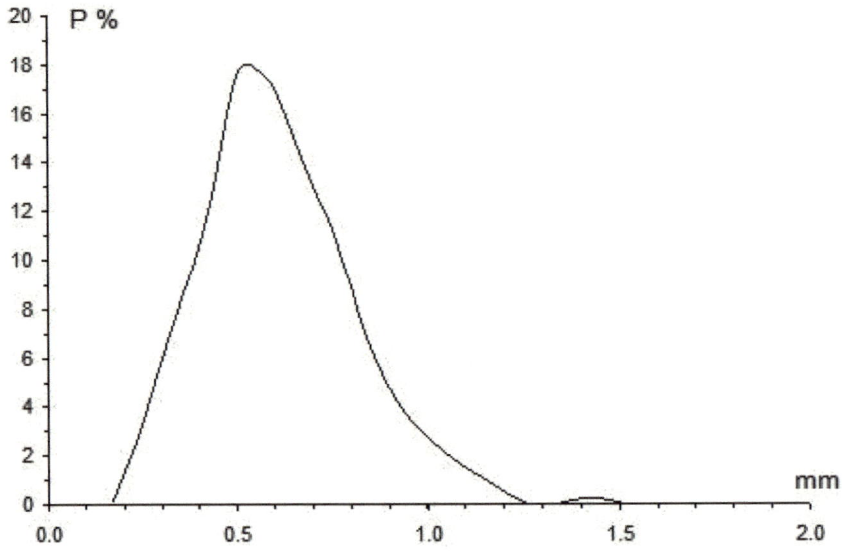

Figure 13.
The distribution of macropores according to their size for the foam concrete sample on the basis of foam stabilized with SiO_2 sol.

Figures 12 and **13** show the distribution of large pores according to their size for the foam concrete samples of average density D_{600}; the study was conducted by means of the electron microscopy.

As it can be seen from the figures, the peak of the pore distribution according to their size in the case of a foam concrete sample based on a stabilized foam is shifted toward a smaller pore diameter (D_{av} = 520 μm). The number of such pores is 18%, the half-width of the peak is 0.44 mm. For the control sample, the peak corresponds to D_{av} = 600 μm, the number of pores of the average diameter is 15%, and the half-width of the peak is 0.52 mm.

Thus, it can be concluded that the stabilization of foam with SiO_2 sol influences both the micro- and macropores of foam concrete. This increase in the foam concrete pore dispersion during the stabilization of foam explains the decrease in its thermal conductivity (**Table 5**).

6. Production of foam concrete on the basis of the stabilized foam in industrial conditions

The results obtained by our group were tested in the industrial conditions in accordance with the technological specifications for the production of foam concrete by cutting technology. In this technology the cutting process begins when a foam concrete mass develops a certain cutting strength. The time of cutting strength attainment can be 17–20 h depending on the composition of the foam concrete. To reduce the time of this technological stage is important from an economic point of view besides that it can improve the productivity of the enterprise.

The studies of samples obtained on the basis of the foam stabilized with SiO_2 sol in production conditions also showed that the foam was not destroyed by the introduction of NaCl additives. The compressive strength of foam concrete products increases up to 38%, and the coefficient of thermal conductivity decreases and corresponds to the foam concrete the class of which is one class lower by the average density. At the same time, the duration of the cutting strength attainment reduces by 7 hours, which significantly speeds up the technological process.

In addition, the shrinkage value of the foam concrete samples was estimated in drying. It was stated to decrease by 18% compared to the control sample, as shown in **Table 9**.

Class by average density	Additive	The time of cutting strength attainment, hour	Thermal conductivity, λ, W/(m•°C)/%	Shrinkage in drying, mm/m
D_{500}	—	17	0.117/100	3.4
	NaCl	10	0.094/80	2.8

Table 9.
Physical and technical characteristics of industrial samples of non-autoclaved foam concrete of average density D_{500}.

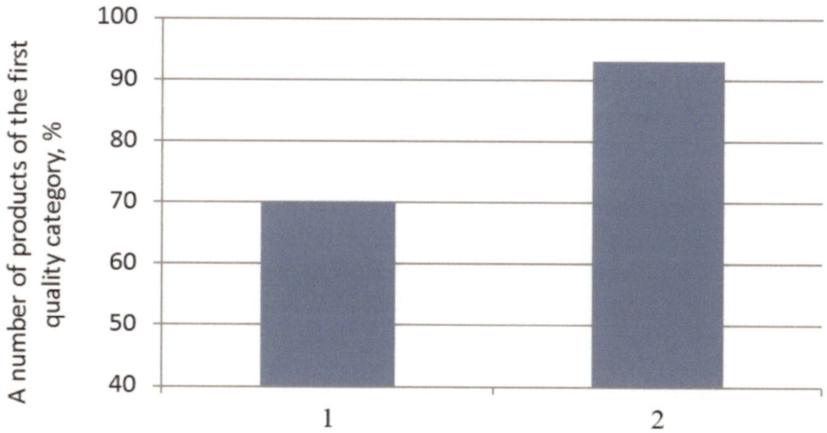

Figure 14.
The number of concrete products of the first quality category: (1) control products, (2) products based on the stabilized foam and NaCl additives.

Figure 15.
Foam concrete products based on the stabilized foam and NaCl additives.

Figure 14 shows the number of foam concrete products of the first quality category which increases by 23% compared to the control products made of foam concrete. These figures confirmed the earlier assumptions.

The quality category of products was evaluated in accordance with the specifications (GOST 21520-89). According to its requirements for the blocks of the first quality category the limit deviations in height can be ±1 mm, in length, thickness and deviation from the rectangular shape it can be ±2 mm. Damage to the corners on one block with a depth of no more than 5 mm, damage to the ribs on one block with a total length of no more than twice the length of the longitudinal edge and a depth of no more than 5 mm are limited by the specifications.

Figure 15 shows the industrial samples of foam concrete products of average density D_{500}.

7. Conclusions

1. For the first time it is proposed to use SiO_2 and $Fe(OH)_3$ sols to stabilize the construction foam on a protein basis in order to obtain high-quality non-autoclaved foam concrete.

2. It is shown that the stability of the foam stabilized with SiO_2 and $Fe(OH)_3$ sols increases up to four times and the coefficient of resistance of the foam in the cement paste increases from 0.9 to 0.98.

3. The method of IR Fourier spectroscopy shows that the mechanism of the foam stabilization is connected with the formation of various chemical bonds between the molecules of the protein foaming agent and sols: hydrogenous in the case of using SiO_2 sol and covalent in the case of $Fe(OH)_3$ sol.

4. It is established that the new chemical bonds result in the formation of spatial stabilizing complexes in the foam film. This leads to an increase in its thickness by one order, which is confirmed by electron microscopy.

5. Heat-insulating foam concrete of average density D_{200} with thermal conductivity coefficient 0.04 W/(m•°C) was obtained on the basis of the foam stabilized with SiO_2 sol.

6. On the basis of the foam stabilized with SiO_2 and $Fe(OH)_3$ sols and NaCl additive, high-quality foam concrete of medium density D_{400}–D_{600} was obtained. Compressive strength increases up to 50%, tensile strength in bending increases up to 69%, shrinkage in drying decreases by 18%, the coefficient of thermal conductivity decreases by 16%, and frost resistance increases by 20 cycles.

7. Studies of the porous structure of foam concrete based on the stabilized foam have shown that the specific surface area of the pores increases twice and the average pore diameter decreases from 600 to 520 μm.

8. Industrial testing of the developed foam concrete of the average density D_{500} has been made. The use of the foam stabilized with SiO_2 sol and the additive of NaCl in its composition allows to reduce the time of cutting strength attainment by 7 hours and to increase the number of products of the first quality category by 23%.

Author details

Anastasia Sychova[1]*, Larisa Svatovskaya[2] and Maxim Sychov[3]

1 Military Space Academy named after A.F. Mozhaisky, St. Petersburg, Russia

2 Department of "Engineering Chemistry and Natural Science", Emperor Alexander I St. Petersburg State Transport University, St. Petersburg, Russia

3 Department of Materials Science, Saint-Petersburg State Institute of Technology, St. Petersburg, Russia

*Address all correspondence to: amsychova@yandex.ru

References

[1] Guo Z, Xu C, Lu Y, Zhou L. Foamability and stability of foam and means of evaluating. No. 4. Chemical Engineering. 2006. 51 p. DOI: 10.1007/s12613-017-1397-2

[2] Sritam S, Indu Siva R, Khwairakpam S. State of the art review on the characteristics of surfactants and foam from foam concrete perspective. Journal of The Institution of Engineers (India). 2018:391-405. DOI: 10.1007/s40030-018-0288-5

[3] Barnes R. Foamed concrete: Application and specification. In: Proceedings of the International Conference on Concrete Construction. London: Kingston University; 2009. pp. 3-9

[4] Indu Siva R, Ramamurthy K. Relative assessment of density and stability of foam produced with four synthetic surfactants. Materials and Structures. 2010:1317-1325. DOI: 10.1617/s11527-010-9582-z

[5] Khitrov A. Technology and properties of foam concrete with the consideration of the injected foam nature [thesis]. Saint-Petersburg; 2006

[6] Makhambetova U, Sultanbekov T, Estemesov Z. Modern Foam Concrete. Saint-Petersburg: Emperor Alexander I St. Petersburg State Transport University; 1997. 159 p

[7] Tikhomirov V. Foams. Theory and Practice of their Production and Destruction. Moscow: Chemistry; 1983. 264 p

[8] Shakhova L, Balyasnikov V. Foaming Agents for Cellular Concrete. Belgorod; 2002. 147 p

[9] Svatovskaya L, Sychova A, Eliseeva N, Bojkova T. The chemical engineering of the sol-gel nonavtoclave foam concrete. In: 17 Internationale Baustofftagung; 23-26 September 2009; F.A. Finger-Institut fur Baustoffkunde Bauhaus-Universitat Weimar Bundesrespublik Deutschland. Tagungsbericht. 2009. Das Band 2. pp. 1203-1206

[10] Kirsanova N, Kovalevich O, Salisheva O, Holohonova L, Rozalenok N, Gelfman M. Workshop on Colloid Chemistry. Saint-Petersburg: Lan; 2005. 256 p

[11] Mikulski V, Kupriyanov V. Building Materials (Materials Science). 2nd ed. Moscow: Publishing House of Association of Construction Universities; 2004. 536 p

[12] Sychova A, Solomahin A, Kotovich V, Svatovskaya L, Kamenev Y. Improving of the Monolithic Foamconcrete Quality for Used in the High-Rise Constructionsd E3S Web of Conferences 33; 2018. DOI: 10.1051/e3sconf/20183302058

[13] Tang F, Xiao Z, Tang J, Jiang L. The effect of SiO_2 particles upon stabilization of foam. Journal of Colloid and Interface Science. 1989;**2**:498. DOI: 10.1007/s12613-017-1397-2

[14] Liu Q, Zhang S, Sun D, Xu J. Aqueous foams stabilized by hexylamine-modified laponite particles. Colloids and Surface A. 2009;**1-3**:40. DOI: 10.1007/s12613-017-1397-2

[15] Alargova R, Warhadpande D, Paunov V, Velev O. Foam superstabilization by polymer microrods. Langmuir. 2004;**24**:10371. DOI: 10.1007/s12613-017-1397-2

[16] Nakayama S, Hamasaki S, Ueno K, Mochizuki M, Yusa S, Nakamura Y, et al. Foams stabilized with solid particles carrying stimuli-responsive

polymer hairs. Soft Matter. 2016:4794. DOI: 10.1039/c6sm00425c

[17] Dong X, Xu J, Cao C, Sun D, Jiang X. Aqueous foam stabilized by hydrophobically modified silica particles and liquid paraffin droplets. Colloids and Surfaces, A: Physicochemical and Engineering Aspects. 2010;**2-3**:181. DOI: 10.1016/j.colsurfa.2009.04.039

[18] Binks B, Horozov T. Aqueous foams stabilized solely by silica nanoparticles. Angewandte Chemie International Edition. 2005;**24**:3722. DOI: 10.1002/ange.200462470

[19] Zhu Q, Zhou H, Song Y, Chang Z, Li W. Modification and investigation of silica particles as a foam stabilizer. International Journal of Minerals. Metallurgy and Materials. 2017:208-215. DOI: 10.1007/s12613-017-1397-2

[20] Kaptay G. Interfacial criteria for stabilization of liquid foams by solid particles. Colloids Surface A. 2003. 67 p: 1-3. DOI: 10.1016/j.colsurfa.2003.09.016

[21] Xing N-L, Yan X-L, Xiang C, Yuan L, Xue L-F. Foam stability in gas injection foaming process. Journal of Materials Science. 2010:6481-6493. DOI: 10.1007/s10853-010-4736

[22] Basnet B, Sarkar N, Park J-G, Mazumder S, Kim I-J. Al_2O_3–TiO_2/ZrO_2–SiO_2 based porous ceramics from particle-stabilized wet foam. Journal of Advanced Ceramics. 2017:129-138. DOI: 10.1007/s40145-017-0225-5

[23] Shakhmenko G, Juhnevica I, Korjakins A. Influence of sol-gel nanosilica on hardening processes and physical-mechanical properties of the cement paste. Procedia Engineering. 2013:1013-1021. DOI: 10.1016/j.proeng.2013.04.128

[24] Nazari A, Riahi S. Splitting tensile strength of concrete using ground granulated blast furnace slag and SiO_2 nanoparticles as binder. Energy and Buildings. 2011:864-872. DOI: 10.1016/j.enbuild.2010.12.006

[25] Nazari A, Riahi S. Abrasion resistance of concrete containing SiO_2 and Al_2O_3 nanoparticles in different curing media. Energy and Buildings. 2011:2939-2946. DOI: 10.1016/j.enbuild.2011.07.022

[26] Sychova A, Svatovskaya L, Starchukov D, Soloviova V, Gravit M. The improving of the concrete quality in a monolithic clip. Magazine of Civil Engineering. 2018;**04**:3-14. DOI: 10.18720/MCE.80.1

[27] Svatovskaya L, Sychova A, Sychov M. Silica sol properties for construction. MATEC Web of Conferences. 2018;**193**. DOI: 10.1051/matecconf/201819303034

[28] Sychova A, Svatovskaya L, Starchukov D, Gera V, Sychov M. Concrete surfacing using surface energy of nanoparticles. International Journal of Engineering & Technology. 2018;**7** (4.7):356-360

[29] Zimon A. Colloid and Surface Chemistry. Moscow: Agar; 2007. 344 p

[30] Shabanova N, Sarkisov P. Fundamentals of Sol-Gel Technology of Nano Silicon Oxide. Moscow: IKC "Akadembook"; 2004. 207 p

[31] Sycheva A, Stepanova I, Eliseeva N, Starchukov D, Solovev D. Nanoadditions in Compositions from the Inorganic Knitting. The Monography. Saint-Petersburg: Emperor Alexander I St. Petersburg State Transport University; 2010. 80 p

Chapter 7

CO_2 Foam for Enhanced Oil Recovery Applications

Ahmed Farid Ibrahim and Hisham A. Nasr-El-Din

Abstract

CO_2-foam yields improved sweep efficiency in enhanced oil recovery (EOR) applications over that of polymers to avoid potential polymer-induced formation damage. In addition to carbon sequestration in underground formations, CO_2 foam has low water content, which also reduces formation damage in water-sensitive formations and allows for fast cleanup. However, foam stability diminishes in harsh environments such as those with high salinity and temperature and when in contact with crude oil. This chapter highlights the different foam-generation mechanisms and the deterioration effect of crude oil on CO_2-foam stability. More specifically, this chapter investigates using nanoparticles and viscosifiers to improve foam stability. Further, the effects of different nanoparticles, including aluminum oxide, copper oxide, and low-cost nanoparticles such as silicon dioxide, will be demonstrated. Field applications of viscoelastic surfactants and polymers in foam systems are also reviewed. The controlling factor for these different systems is the foam stability and improved oil recovery.

Keywords: CO_2 foam, crude oil effect, foam stability, EOR, nanoparticles, viscoelastic surfactants

1. Introduction

Gas injection has been used widely for enhanced oil recovery (EOR) application. Carbon dioxide (CO_2) gas injection was started for EOR applications in the 1950s to improve oil recovery and provide for carbon sequestration in underground formations [1, 2]. However, CO_2 injection proved impractical at that time due to its low viscosity compared to formation fluids, leading to viscous fingering and early breakthrough [3]. Hence, the sweep efficiency and recovery factor were low [4]. Polymer flooding was introduced to reduce the mobility ratio of the displacing fluid to the displaced fluid [5]. The mobility ratio can be calculated using Eq. (1):

$$M = \frac{\lambda_{displacing}}{\lambda_{displaced}} = \frac{(k_e/\mu)_{displacing}}{(k_e/\mu)_{displaced}}, \quad (1)$$

where M is the mobility ratio of the displacing fluid to the displaced fluid, λ is the fluid mobility, k_e is the effective fluid permeability, and μ is the fluid viscosity. Using polymer flooding increases the displacing fluid viscosity. As a result, the mobility ratio decreases, and the viscous fingering is reduced. Hence, sweep efficiency improved, as shown in **Figure 1**. However, polymer flooding is associated with formation damage due to physical adsorption of the high-molecular-weight

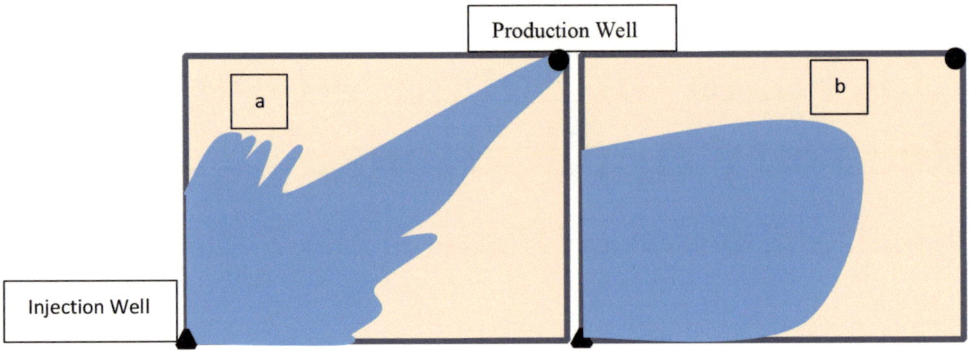

Figure 1.
Areal sweep efficiency, (a) low sweep efficiency and early breakthrough due to viscous fingering at unfavorable mobility ratio values (M < 1) and (b) high sweep efficiency at favorable mobility ratio (M > 1).

polymer on the rock surface and mechanical trapping within the smaller-diameter pore throats [6, 7].

CO_2 foam was introduced in the 1960s as a replacement for polymers to avoid formation damage [8]. Foam has low water content, which reduces formation damage in water-sensitive formations and allows fast cleanup [9]. Foam is a dispersion of a gas (nitrogen, carbon dioxide, or methane) as a non-wetting fluid in a continuous wetting phase. The wetting phase is water that contains surfactant at a particular concentration that is above the critical micelle concentration (CMC). The liquid film separates the gas phase from each other, the outer membranes of the gas bubbles, called *foam lamella*e. The first surfactant families selected for EOR method were petroleum and synthetic aromatic sulfonates [such as alpha-olefin sulfonate (AOS)] because of their availability, lower adsorption on porous rocks, high compatibility with hard water, and good wetting and foaming properties [10, 11].

Bulk foam can be characterized by several properties such as quality, texture, stability, and foam density [12]. Foam quality is the volume percent of gas within foam at a specified pressure and temperature [13]. Foam quality for EOR applications is typically 75–90%. Foam texture is a measure of the average gas bubble size. Foam stability depends on the chemical and physical properties of the surfactant-stabilized water film separating the gas bubbles (lamellae). Foams are metastable systems; accordingly, all foams will eventually break down. Foam stability is measured by the half-life time, which is the time required to lose 50% of the foam volume [14]. In general, as a foam texture becomes finer, the foam will be more stable and will have greater resistance to flow in matrix rock. Foam flow resistance in porous media is measured by the mobility reduction factor (MRF). MRF is defined as the ratio of total mobility of CO_2/brine to foam mobility. When foams become more stable, more resistance to flow is expected and leads to a higher mobility reduction [15].

2. Foam generation

Foam generates in porous media through three different mechanisms: (a) snap-off, (b) lamella division, and (c) leave-behind [16, 17].

In the snap-off mechanism, lamellae are created in gas-filled pore throats as a result of the capillary pressure difference between the pore body and the pore neck [16]. **Figure 2a** shows the foam-generation process by the snap-off mechanism. As the gas phase displaces the liquid zone, the difference in capillary pressure between the pore body and the pore throat forces the wetting phase (water) to flow back and then snap-off the gas phase.

Lamella division generally occurs when a foam lamella that is larger than that of the pore body approaches a "branching point" and divides into two or more bubbles (**Figure 2b**). If the lamella is at a branching point with more than one path that requires the same pressure for the lamella to flow, the lamella divides into two bubbles or lamellae [18].

Leave-behind occurs when the gas enters a porous medium that is initially saturated with a liquid or when two gas fronts approach a pore space that is filled with liquid; these processes squeeze the liquid into a lamella (**Figure 2c**). The leave-behind mechanism typically forms a weak foam because the generated lamella is parallel to the flow direction [17, 18].

3. Experimental evaluation methods

3.1 Bulk foam stability and microscopic analysis

Foams are generally described in terms of their foamability, which is the ability of a foaming solution (water in the presence of foaming agents) to form a foam. Bulk foam stability or formability tests are static tests that can be used for screening different parameters such as foaming agents, concentrations, salinity, and effect of crude oil [20]. At ambient conditions, foamability of a solution can be studied by performing a shake test [21]. At high-pressure/high-temperature (HP/HT) conditions, foam can be generated by gas sparging into an HP/HT

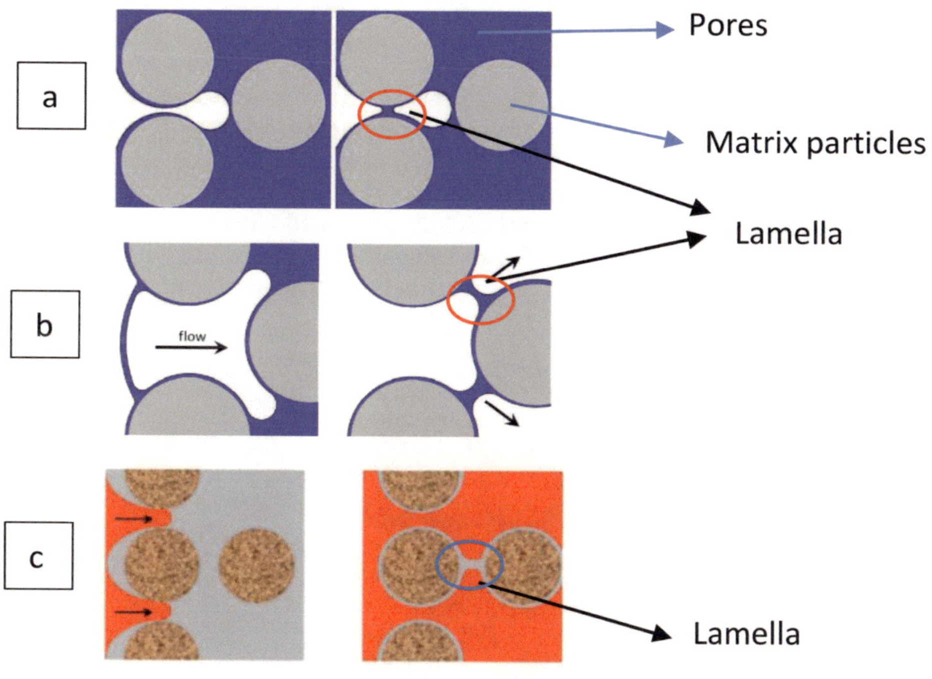

Figure 2.
Foam lamella generation mechanisms: (a) snap-off mechanism, (b) lamella division, and (c) leave-behind (after [19]).

visual graduated cell [22, 23]. The foam half-life time can be measured by monitoring the foam height over time.

Additionally, bubble-scale experiments can be conducted under an optical microscope to investigate the foam stability [24]. The foam bubbles are allowed to stabilize and are then placed on a microscope slide. The foam texture and the thin liquid films (lamellae) are then monitored with time to investigate the foam decay rate [23]. **Figure 3** shows a microscopic image for an AOS foam system in contact with a crude oil.

3.2 Macroscopic sweep experiments

Different flood experiments can be conducted to evaluate the foam performance in porous media. Glass bead packs or cores can be used to represent the porous media. Three distinctly different modes are used for foam injecting: (1) alternative injection of gas and liquid with foaming agents, (2) co-injection of the gas and the liquid phase at the same time, and (3) injection of pregenerated foam. Foam stability usually quantified by the oil recovery and MRF. The MRF can be calculated by comparing the pressure drop across the core during foam injection to the pressure drop after gas injection [15], as described in Eq. 2:

$$MRF = \frac{\mu_f}{\mu_b} = \frac{\left[\frac{kA\Delta p}{QL}\right]_f}{\left[\frac{kA\Delta p}{QL}\right]_b} = \frac{\Delta p_f}{\Delta p_b}, \tag{2}$$

where Q is the total flow rate, k is the absolute core permeability, A is the cross-section area of the core, L is the core length, μ is the viscosity, Δp is the pressure drop across the core, and the subscripts "f" and "b" represent the experiments with and without foam, respectively. The pressure buildup along the porous medium indicates foam-generation and gas mobility reductions [25]. A higher pressure drop signifies viscous foam and considerable resistance to gas mobility in porous media.

Dual coreflood experiments can be conducted to evaluate the divergent ability of foam systems within heterogeneous systems [23]. The foam is injected in two parallel cores with different permeabilities. The stable foam will be generated in the high-permeability formation and divert the flow toward the low-permeability core that improves the sweep efficiency and increases the oil recovery.

Macroscopic sweep experiments are usually combined with X-ray computed tomography (CT) measurements. CT scan analysis can be used to determine the porosity, oil distribution, and foam propagation inside the porous medium.

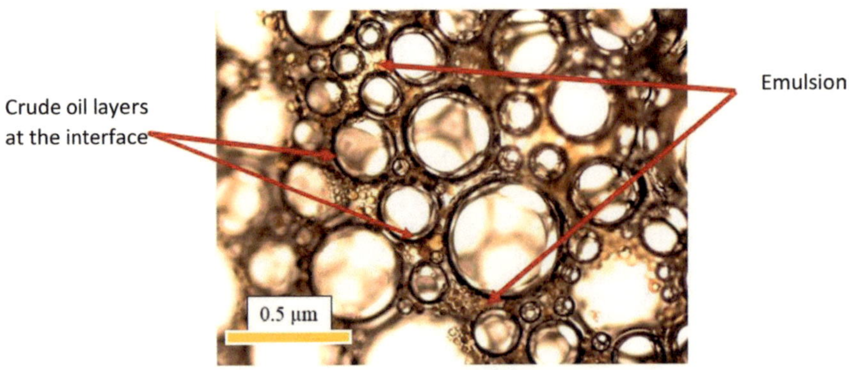

Figure 3.
Microscopic image of AOS foam in contact with a crude oil (5×) [23].

4. Deterioration effect of crude oil

The instability effect of crude oil on the CO_2-foam system is another challenge for the use of this foam in EOR applications [26]. Crude oil composition, especially the presence of light components, decreases foam stability [27]. Foam stability decreases in contact with crude oil as a result of direct surface interactions between oil and foam. These interactions are governed by three main mechanisms: entry of oil droplets into the gas–liquid interface, spreading of oil on the gas–liquid interface, and formation of an unstable bridge across lamellae [27–30]. These three mechanisms can be quantified as a function of the interfacial tensions between oil, gas, and water by evaluating the entering coefficient (E), spreading coefficient (S), and bridging coefficient (B) [26]. E, S, and B can be calculated as follows (Eqs. 3–5):

$$E = \sigma_{gw} + \sigma_{ow} - \sigma_{go} \tag{3}$$

$$S = \sigma_{gw} - \sigma_{ow} - \sigma_{go} \tag{4}$$

$$B = \sigma_{gw}^2 + \sigma_{ow}^2 - \sigma_{go}^2 \tag{5}$$

where σ_{gw}, σ_{ow}, and σ_{go} are the interfacal tensions between CO_2 and water, oil and water, and oil and CO_2, respectively. **Figure 4** presents a flowchart to predict the foam stability when in contact with oil, as indicated by the E, S, and B coefficients [27]. The oil droplets should be able to enter the gas-water interface to destabilize the foam. Once the entry condition is achieved (E is positive) and the oil droplets spread on the gas–liquid interface (S is positive), the gas/water interface will expand. As a result, the foam lamellae become thin and rupture, thus weakening the foam. If there is no spreading (S is negative) and the oil droplets form an emulsion at the gas/water interface, the foam film may rupture once oil droplets bridge between both surfaces of the lamellae (B is positive).

Ibrahim and Nasr-El-Din [23, 24] found that the AOS foam in contact with oil became unstable and decayed very fast, dissolving completely in 30 min compared

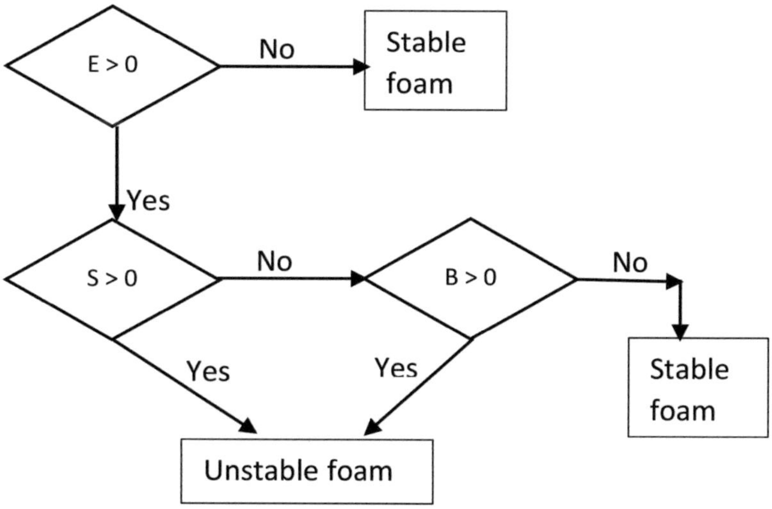

Figure 4.
Flowchart to predict foam stability from E, S, and B coefficients.

to more than 5 h in the absence of oil [24]. Two reasons account for the adverse effect of crude oil on foam decay. First, oil droplets tend to spread along with the gas/liquid interface. As a result, the stable gas/liquid interface becomes unstable, which accelerates the rupture of the foam lamellae. Second, oil forms an emulsion in the foam lamellae. The oil droplets in the unstable emulsion agglomerate and accelerate the drainage of the foam lamellae; hence, the foam decays faster [28].

In **Figure 3**, the crude oil forms layers in the interface between the gas and liquid phases, and an emulsion forms inside the AOS foam lamellas. **Figure 5** shows microscope images for AOS foam lamellae in the presence of oil. The oil droplets in the unstable emulsion agglomerate and accelerate the drainage, where the lamella thickness decreases over time. Hence, the foam becomes unstable and decay faster.

Figure 6 plots the foam height over time that gives indication for the decay rate from a visual cell experiment in which the foam lamellae were in contact with crude oil. At room temperature, the initial AOS foam height was 20 cm; then the foam decayed over time to reach 10 cm after 15 min (foam half-life = 15 min). In contact with oil, the foam decayed faster, and the half-life time decreased to 3 min.

Figure 7a shows the change of the normalized foam half-life as a function of temperature. The normalized half-life time is the half-life time for the foam system divided by the half-life time of AOS system at room temperature without crude oil. With increasing temperature, the foam became unstable and decayed more quickly than at room temperature. The half-life for the AOS system in the absence of crude oil at 150°F also decreased to 0.13 of its value at 77°F. As the temperature increases,

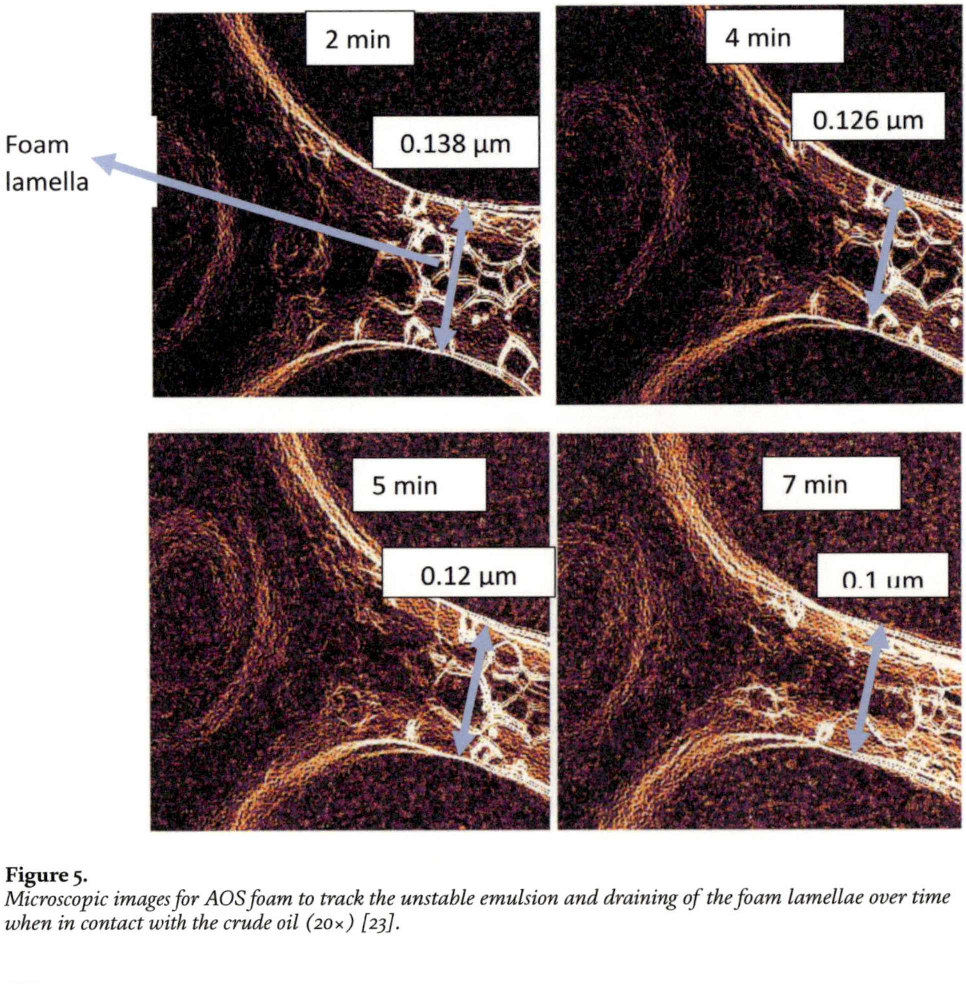

Figure 5.
Microscopic images for AOS foam to track the unstable emulsion and draining of the foam lamellae over time when in contact with the crude oil (20×) [23].

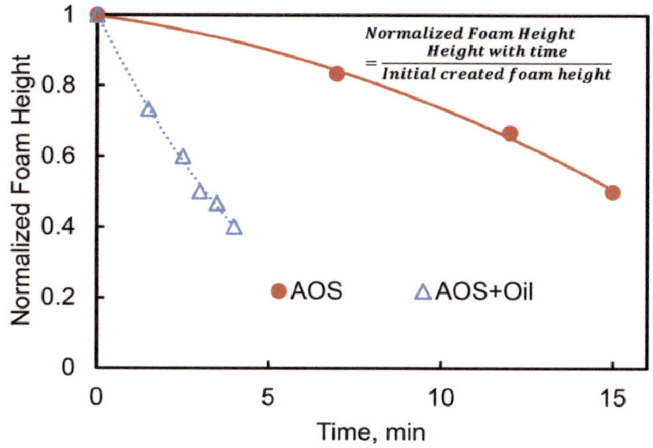

Figure 6.
Normalized foam height for AOS foam system in the absence and presence of crude oil as a function of time at 77°F and 800 psi.

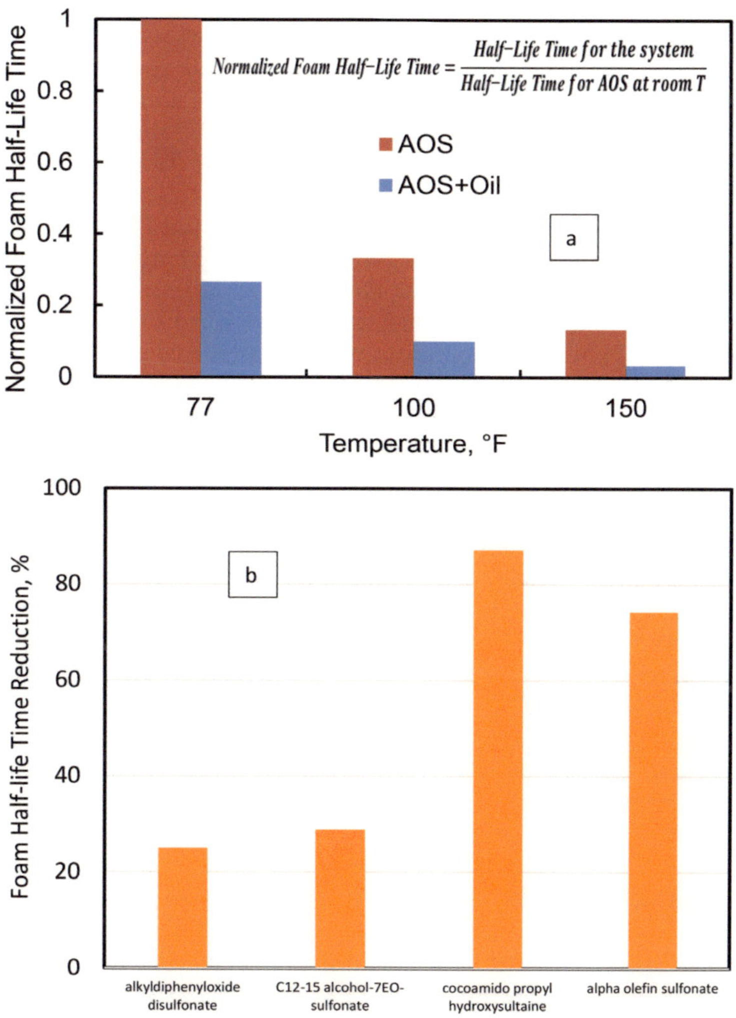

Figure 7.
Effect of 1 vol% crude oil in foam stability, (a) foam stability at different temperatures at 800 psi for 0.5 wt% AOS foam system and (b) foam half-life time reduction percent for different surfactant systems.

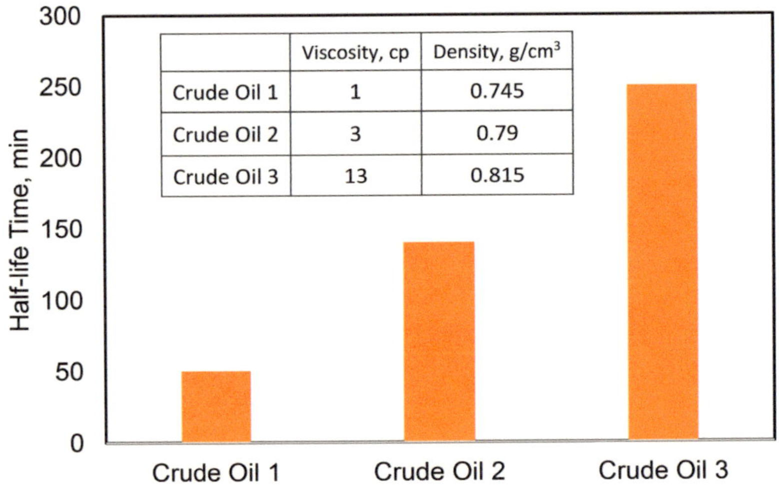

Figure 8.
Effect of oil density and viscosity on foam stability [29].

the liquid viscosity decreases, which leads to faster drainage. As a result, gas bubbles coalesce faster [31]. In contact with crude oil, the AOS was not able to generate stable foam, where the half-life was 3.9 and 0.45 min at 77 and 150°*F*, respectively.

The E, S, and B coefficients for the AOS system at 150°*F* and 800 psi were found to be 7.96, 6.68 mN/m, and 231.74 $(mN/m)^2$. Based on the flowchart in **Figure 4**, the generated foam is not stable. Increasing the pressure to 1200 psi, the values for the three coefficients slightly decreased to 6.67, 5.38 mN/m, and 141.29 $(mN/m)^2$, still greater than zero. A positive E value indicates that the oil-entering condition was initially favorable. However, once the oil entered the foam system as an emulsion on the AOS and AOS/SiO_2 foam lamellae, it started to spread on the surfaces of the foam bubbles (S > 0) and generated unstable bridges (B > 0) that quickly broke the lamellae and destabilized the foam. The microscopic images of the AOS foam system in contact with crude oil in **Figures 3** and **5** confirm this behavior.

Similar results were observed by Simjoo et al. [27]. Adding 1 vol% crude oil to different surfactant systems decreases the foam stability and the half-life time. **Figure 7b** shows the change in the foam half-life time due to the presence of crude oil. Regardless of the surfactant type, adding crude oil to the foam increases the lamella drainage and decreases the foam stability.

The previous experimental results demonstrate the detrimental effect of crude oil on foam stability [29]. However, regardless of the surfactant used, crude oil with higher hydrocarbon chain lengths has a lower effect on foam stability. **Figure 8** shows the half-life of Coco/SDS (a foam system with a 1:1 mixture of cocobetaine and sodium dodecyl sulfate surfactants) in contact with different crude oils with different densities and viscosities at room temperature. These data show that the higher the density and viscosity, the lower the effect of crude oil on foam stability.

5. Improving foam stability using nanoparticles

The previous section shows the instability of foam systems in contact with crude oil at high temperature. Nanoparticles have been examined extensively as a means to stabilize foams used in oil-production operations, including those in high-salinity and high-temperature environments [9, 20–25, 32, 33]. This behavior is due to the nanoparticles' adsorption to the interface between the gas and liquid phases and minimizes the contact area between them; as a result, they can build a strong barrier that prevents bubble coalescence. **Figure 9a** shows a microscopic image for SDS foam system that was stabilized with Al_2O_3 nanoparticles [32]. It shows the nanoparticle adsorption at the lamella surface. Hence, the nanoparticle-stabilized foams are expected to be durable and highly resistant to unfavorable reservoir conditions including high salinity, high temperatures, and the presence of crude oil. Silica nanoparticles are currently regarded as most effective for improving foam stability [20–25].

Nanoparticles' size greatly affect the foam stability. Different experimental results for the usage of silica nanoparticles showed that the smaller the nanoparticle size, the higher the foam stability [32]. The small particle will move faster to the gas–liquid interface compared to the larger nanoparticle size. Hence, the nanoparticle adsorption and concentration in the lamella surface increase and the foam become more stable. This behavior of foam stability with nanoparticle size greatly depends on foam quality, salinity, and nanoparticle hydrophobicity. Larger-size nanoparticles improve the foam stability at foam quality of 70–80%, while smaller size nanoparticles improve the foam stability at quality of 50–60% [9]. In addition, 140 nm silica nanoparticles with contact angle of 86° increased the foam stability greater than 100 nm silica nanoparticles with contact angle of 54° [21].

Emrani et al. [21] studied the effect of adding 140 nm silica nanoparticles to AOS foam system. This work achieved a stable foam with an MRF of 8, which was four

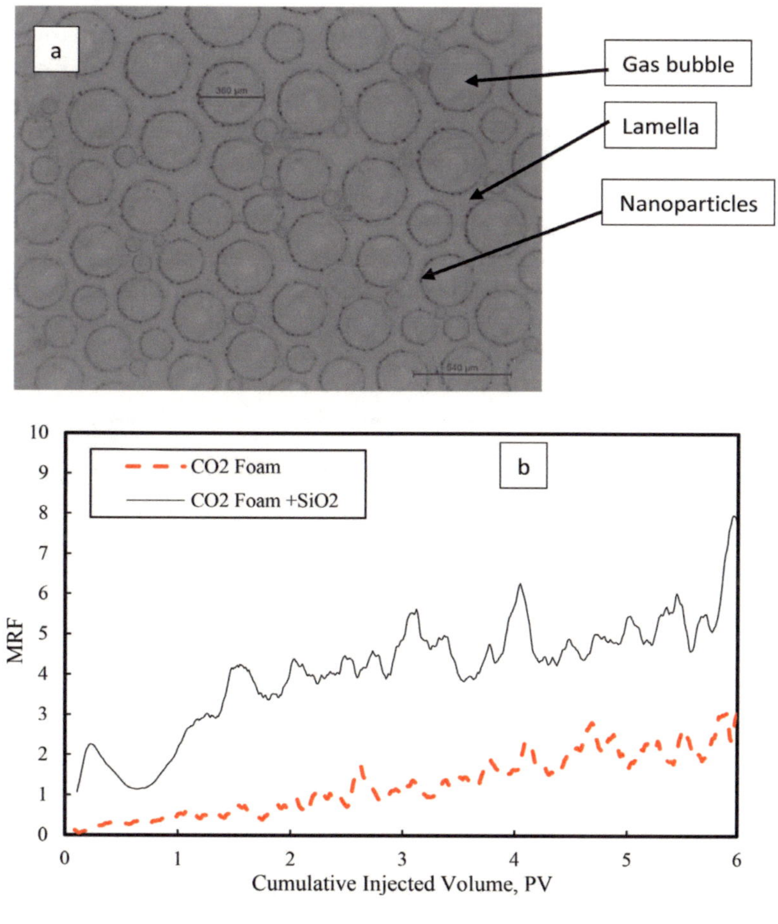

Figure 9.
Foam stability improvement by nanoparticles, (a) microscopic image for SDS/Al_2O_3 foam system [32] and (b) MRF for 0.5 wt% AOS solutions in the absence and presence of 0.1 wt% nanoparticles at 77°F [21].

times the MRF for AOS foam system. Adding nanoparticles creates a fine texture foam that increases the apparent viscosity and the MRF (**Figure 9b**). Similarly, in a bulk stability test, adding nanoparticles to the AOS foam system tripled the half-life of its original value without nanoparticles [24]. In addition, nanoparticles are adsorb on the interface between the gas and liquid phases, creating thick, solid films that provide a barrier to film thinning and inter-bubble diffusion. Hence, in the presence of crude oil, the spreading of the oil droplets along the foam lamellae decreases and prevents bridge formation. Interfacial tension (IFT) measurements for silica nanoparticle foam system showed that the spreading and bridging coefficients have negative values (−0.69, −4.43 at 1200 psi and 150°F), thus evidencing improved foam stability.

Nanoparticle concentration in the interface between the liquid and gas phases is a critical parameter and should increase to a certain threshold to stabilize the foam [34]. **Figure 10** shows the change in foam half-life time with increasing silica nanoparticle concentrations. At low concentrations, the liquid/gas interface is not saturated, and low foam stability is generated. With increased nanoparticle concentration, foam stability increases. However, at higher concentration, nanoparticles agglomerate and form bigger particles that negatively impact the foam stability. Zeta potential measurements in **Figure 10** show a reduction of the absolute zeta potential value with increasing nanoparticle concentration from 0.1 to 0.2 wt%, which indicates a stable suspension. At nanoparticles' concentration higher than 0.2 wt%, nanoparticles become unstable and agglomerate which is indicated by increasing the zeta potential value.

Coreflood experiments by Ibrahim and Nasr-El-Din [23, 24] compared EOR results using an AOS foam system versus a silica-nanoparticle system. **Figure 11** shows total oil recovery after tertiary recovery using different foam systems. A water-assisted gas (WAG) system was able to increase the oil recovery to 60%. AOS generated a weak foam with a similar MRF to that of the CO_2/water system, and oil recovery increased by 1.8%. Adding silica nanoparticles to the foam system increased oil recovery to 68.2%.

In a high-salinity environment, the absolute zeta potential for suspensions will decrease [35]. As a result, nanoparticles will have a high affinity to agglomerate. To prevent the instability of nanoparticles in a high-salinity environment, surface-modified nanoparticles were used to provide steric repulsion between particles.

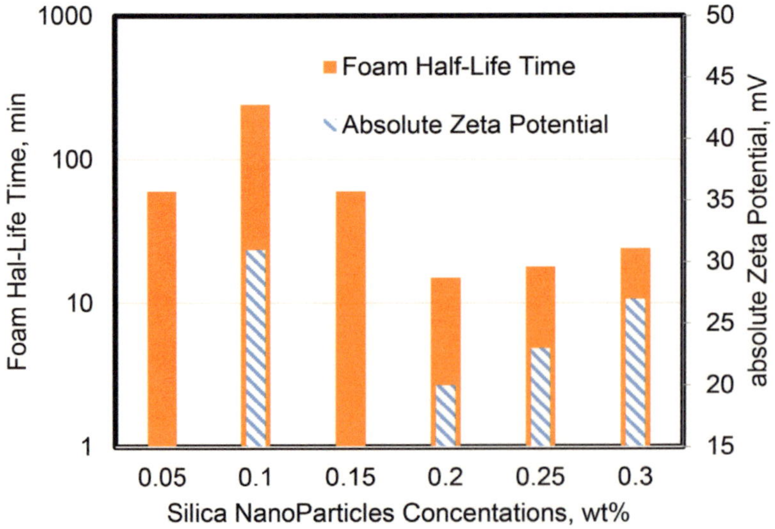

Figure 10.
Effect of silica nanoparticle concentrations on foam stability.

Nanoparticle surface wettability is another critical parameter of foam stability. Generally, nanoparticle surfaces should be hydrophilic enough to disperse in water but hydrophobic enough to accumulate at the interface between the water and gas. Nanoparticles coated with 50% SiOH dichlorodimethylsilane generated a stable foam compared to polyethylene glycol (PEG)-coated nanoparticles or dichloro-dimethylsilane-coated nanoparticles with higher SiOH% even at 8 wt% NaCl salt concentration [36].

In addition to silica, other types of nanoparticles can also be used to improve foam stability. About 0.1 wt% Fe_2O_3 nanoparticles were able to increase the AOS foam half-life time from 1 to 7 h at 75°F and 300 psi [37]. However, these nanoparticles tend to aggregate due to their large surface area, which is confirmed by low absolute zeta potential values. An experimental work by Bayat et al. [38] compared the foam stability using SiO_2, Al_2O_3, TiO_2, and CuO nanoparticles. They found that the optimum concentration for these nanoparticles was 0.008 wt%. **Figure 12** shows the recovery factor and foam half-life for the four different systems. SiO_2 foam had the highest foam stability and oil recovery compared to the other nanoparticles.

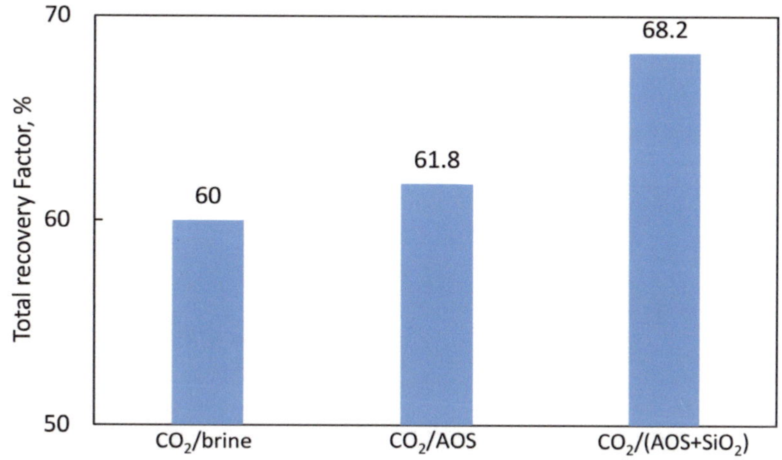

Figure 11.
Total oil recovery for different foam systems with adding 0.5 wt% AOS surfactant and 0.1 wt% silica nanoparticles.

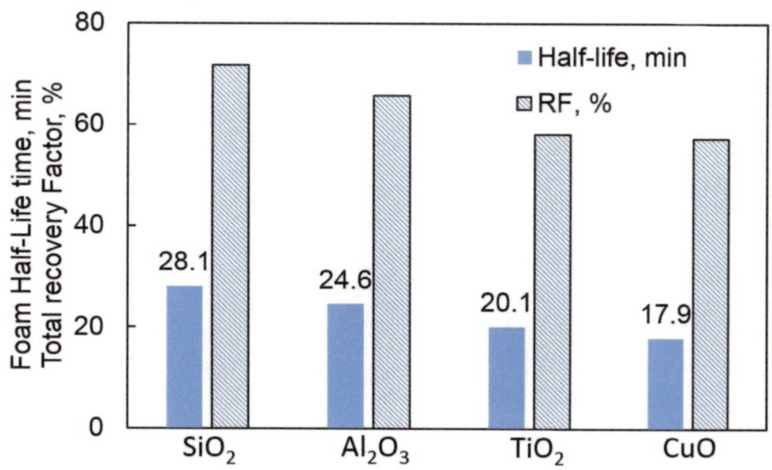

Figure 12.
Effect of nanoparticle type on the foam stability and oil recovery factor.

6. Improving foam stability using viscosifiers

The addition of thickeners such as polymers or viscoelastic surfactant (VES) to CO_2 foam improves foam stability by increasing lamella viscosity that delays

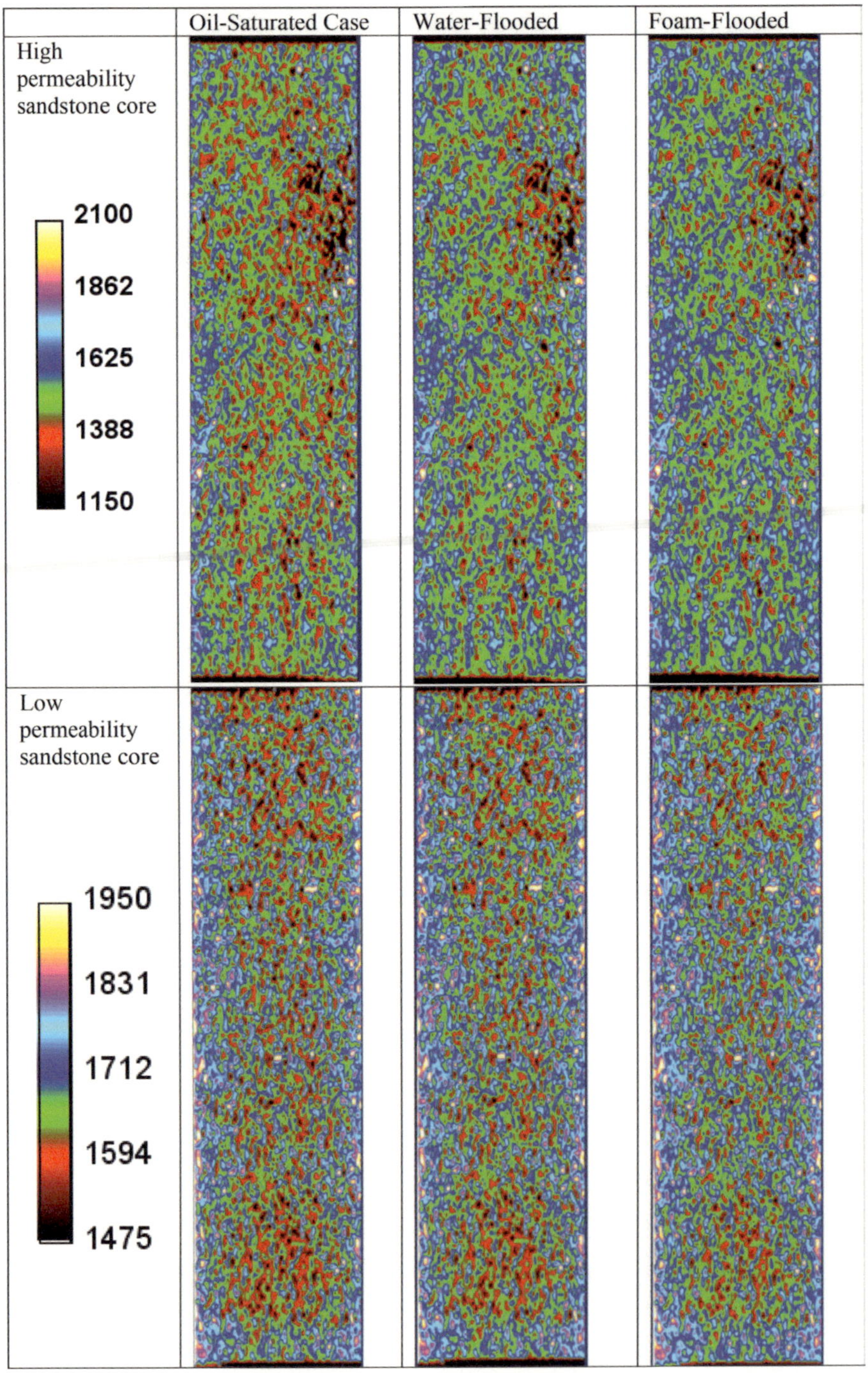

Figure 13.
CT scan scans for Boise and Berea cores during the dual-core flood experiment after oil saturation, waterflooding, and foam-flooding stages.

the lamella drainage effect at high temperature and in contact with crude oils [23, 33, 39, 40].

A dual coreflood experiment was conducted by Ibrahim and Nasr-El-Din [23, 24] to investigate the divergent ability of VES-stabilized foam in heterogeneous formations (with permeability contrast of 700/100). The VES was able to increase the foam stability and improve the sweep efficiency. The oil recovery after the waterflooding stage was 19 and 55% from the low and the high-permeability cores, respectively. Low sweep efficiency in the low-permeability cores was found where the residual oil saturation was 36 vol% compared to 31 vol% in the high-permeability core. When injecting VES into the foam system, the sweep efficiency improved, and the residual oil saturation decreased to 23 and 27 vol% in the low- and the high-permeability cores, respectively. The total oil recovery after foam injection was found to be 50 and 62 vol% of the original oil in place (OOIP).

Figure 13 shows oil saturation distribution along the two cores after the foam-flooding stage compared to the oil-saturated and the waterflooded cases. For the high-permeability core, most of the oil was produced during waterflooding, and the recovery factor increased only by 10% after foam flooding. The slight change in distribution of the red areas in **Figure 13** indicates oil recovery from the waterflooded and the foam-flooded cases. In the case of low-permeability core, most of the reduction in the red color distribution happened after foam flooding.

Figure 14 shows the results of using polymers as a thickener to improve the foam stability [41]. Wei et al. [41] used xanthan gum (molecular weight of $50–100 \times 10^4$) to improve the foam stability for a sulfobetaine-based surfactant foam system at 90°C and 1450 psi. As the polymer concentration increases, the liquid phase viscosity increases. As a result, lamella drainage decreases, and foam-stability half-life increases. However, an increase in polymer concentration also increases the surface-tension forces and thus decreases the system foamability. Stable foam with a higher half-life increases the apparent viscosity for the displacing fluid that improves the sweep efficiency. The oil recovery for the polymer-foam system was 43.2%, compared to 21.8% in the case of the polymer-free foam system.

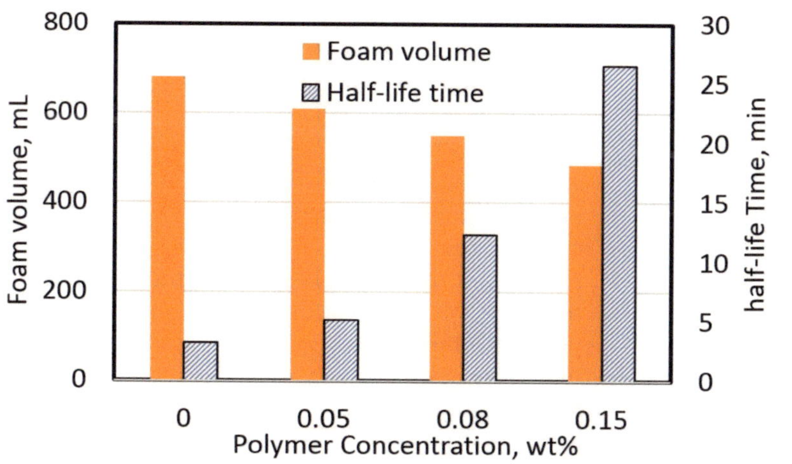

Figure 14.
Effect of polymer concentration on the generated foam volume and its half-life.

7. Summary

This chapter reviews the application of CO_2 foam in enhanced oil recovery and establishes the following main points:

1. Foam lamellae can be generated by a snap-off, lamella division, and leave-behind mechanisms. The foam generated by the leave-behind mechanism is weak because the lamellae are parallel to the flow direction.

2. Crude oil has a negative impact on foam stability, and its deterioration effect can be determined by calculating entering, spreading, and bridging coefficients as a function of oil/gas/water interfacial tension.

3. In harsh environments such as those with high salinity or high temperature, nanoparticles can be used to improve foam stability. SiO_2 with the modified surface was found to be the more effective and popular nanoparticles in foam-stability applications.

4. At high-temperature conditions, VES can be used as a thickener to decrease the rates of lamella drainage and foam decay.

CO_2 foam can be used to improve the oil recovery in EOR process. However, surfactant-based foam provides unstable, and low sweep efficiency will be observed. As a result, nanoparticles or viscosifiers should be used to improve the foam stability.

Acknowledgements

We thank Gia Alexander for proofreading the chapter.

Nomenclature

A	cross-section area of the core
AOS	alpha-olefin sulfonate
B	bridging coefficient
CMC	critical micelle concentration
CT	X-ray computed tomography
E	entering coefficient
EOR	enhanced oil recovery
K	absolute core permeability, md
$k\ e$	effective fluid permeability, md
L	core length
M	mobility of the displacing fluid to the displaced fluid ratio
MRF	mobility reduction factor
PEG	polyethylene glycol
Q	total flow rate
S	spreading coefficient
Δp	pressure drop across the core
λ	fluid mobility, md/cp
μ	fluid viscosity, cp
$\sigma\ gw$, $\sigma\ ow$, and $\sigma\ go$	interfacal tensions between CO_2 and water, oil and water, and oil and CO_2, respectively

Author details

Ahmed Farid Ibrahim* and Hisham A. Nasr-El-Din
Texas A&M University, College Station, TX, USA

*Address all correspondence to: ahmedmfaried@gmail.com

References

[1] Gozalpour F, Ren S, Tohidi B. CO_2 EOR and storage in oil reservoir. Oil & Gas Science and Technology. 2005;**60**(3):537-546

[2] Rezk MG, Foroozesh J, Zivar D, Mumtaz M. CO_2 storage potential during CO2 enhanced oil recovery in sandstone reservoirs. Journal of Natural Gas Science and Engineering. 2019;**66**:233-243

[3] Han J, Lee M, Lee W, Lee Y, Sung W. Effect of gravity segregation on CO_2 sequestration and oil production during CO_2 flooding. Applied Energy. 2016;**161**:85-91

[4] Chen Y, Elhag AS, Poon BM, Cui L, Ma K, Liao SY, et al. Ethoxylated cationic surfactants for CO2 EOR in high temperature, high salinity reservoirs. In: SPE Improved Oil Recovery Symposium; 2012/1/1. Tulsa, Oklahoma, USA: SPE: Society of Petroleum Engineers; 2012. p. 15

[5] Zhang Y, Huang SS, Luo P. Coupling immiscible CO_2 technology and polymer injection to maximize EOR performance for heavy oils. Journal of Canadian Petroleum Technology. 2010;**49**(05):25-33

[6] Dai C. Chapter seven—formation damage during chemical flooding. In: Yuan B, Wood DA, editors. Formation Damage During Improved Oil Recovery. Cambridge, MA: Gulf Professional Publishing; 2018. pp. 275-304

[7] Yuan B, Wood DA. A comprehensive review of formation damage during enhanced oil recovery. Journal of Petroleum Science and Engineering. 2018;**167**:287-299

[8] Bian Y, Penny GS, Sheppard NC. Surfactant formulation evaluation for carbon dioxide foam flooding in heterogeneous sandstone reservoir. In: SPE Improved Oil Recovery Symposium; 2012/1/1. Tulsa, Oklahoma, USA: SPE: Society of Petroleum Engineers; 2012. p. 16

[9] Xiao C, Balasubramanian SN, Clapp LW. Rheology of viscous CO_2 foams stabilized by nanoparticles under high pressure. Industrial & Engineering Chemistry Research. 2017;**56**(29):8340-8348

[10] Negin C, Ali S, Xie Q. Most common surfactants employed in chemical enhanced oil recovery. Petroleum. 2017;**3**(2):197-211

[11] Farajzadeh R, Krastev R, Zitha PLJ. Foam films stabilized with alpha olefin Sulfonate (AOS). Colloids and Surfaces A: Physicochemical and Engineering Aspects. 2008;**324**(1):35-40

[12] Schramm LL. Foams: Fundamentals and Applications in the Petroleum Industry. Washington, DC: American Chemical Society; 1994

[13] Belyadi H, Fathi E, Belyadi F. Chapter five - hydraulic fracturing fluid systems. In: Belyadi H, Fathi E, Belyadi F, editors. Hydraulic Fracturing in Unconventional Reservoirs. Boston: Gulf Professional Publishing; 2017. pp. 49-72

[14] Liu Y, Grigg RB, Bai B. Salinity, pH, and surfactant concentration effects on CO2-foam. In: SPE International Symposium on Oilfield Chemistry; 2005/1/1. The Woodlands, Texas: SPE: Society of Petroleum Engineers; 2005. p. 11

[15] Yin G, Grigg RB, Svec Y. Oil recovery and surfactant adsorption during CO2-foam flooding. In: Offshore Technology Conference; 2009/1/1. Houston, Texas: OTC: Offshore Technology Conference; 2009. p. 14

[16] Falls AH, Hirasaki GJ, Patzek TW, Gauglitz DA, Miller DD, Ratulowski T. Development of a mechanistic foam simulator: The population balance and generation by snap-off. SPE Reservoir Engineering. 1988;**3**(03):884-892

[17] Chen M, Yortsos YC, Rossen WR. Insights on foam generation in porous media from pore-network studies. Colloids and Surfaces A: Physicochemical and Engineering Aspects. 2005;**256**(2):181-189

[18] Emadi A, Sohrabi M, Jamiolahmady M, Irland S, Robertson G. Mechanistic study of improved heavy oil recovery by CO2-foam injection. In: SPE Enhanced Oil Recovery Conference; 2011/1/1. Kuala Lumpur, Malaysia: SPE: Society of Petroleum Engineers; 2011. p. 19

[19] Hematpur H, Mahmood SM, Nasr NH, Elraies KA. Foam flow in porous media: Concepts, models and challenges. Journal of Natural Gas Science and Engineering. 2018;**53**:163-180

[20] Ibrahim AF, Emrani A, Nasr-El-Din H. Stabilized CO_2 foam for EOR applications. In: Carbon Management Technology Conference; 2017/8/24. Houston, Texas, USA: CMTC: Carbon Management Technology Conference; 2017. p. 19

[21] Emrani AS, Ibrahim AF, Nasr-El-Din HA. Mobility control using nanoparticle-stabilized CO_2 foam as a hydraulic fracturing fluid. In: SPE Europec Featured at 79th EAGE Conference and Exhibition; 2017/6/12. Paris, France: SPE: Society of Petroleum Engineers; 2017. p. 16

[22] Singh R, Mohanty KK. Foam flow in a layered, heterogeneous porous medium: A visualization study. Fuel. 2017;**197**:58-69

[23] Farid Ibrahim A, Nasr-El-Din H. An experimental study for the using of nanoparticle/VES stabilized CO2 foam to improve the sweep efficiency in EOR applications. In: SPE Annual Technical Conference and Exhibition; 2018/9/24. Dallas, Texas, USA: SPE: Society of Petroleum Engineers; 2018. p. 15

[24] Farid Ibrahim A, Nasr-El-Din H. Stability improvement of CO2 foam for enhanced oil recovery applications using nanoparticles and viscoelastic surfactants. In: SPE Trinidad and Tobago Section Energy Resources Conference; 2018/6/22. Port of Spain, Trinidad and Tobago: SPE: Society of Petroleum Engineers; 2018. p. 17

[25] Khajehpour M, Etminan SR, Goldman J, Wassmuth F, Bryant S. Nanoparticles as foam stabilizer for steam-foam process. SPE Journal. 2018;**23**(06):2232-2242

[26] Lau HC, O'Brien SM. Effects of spreading and nonspreading oils on foam propagation through porous media. SPE Reservoir Engineering. 1988;**3**(03):983-986

[27] Simjoo M, Rezaei T, Andrianov A, Zitha PLJ. Foam stability in the presence of oil: Effect of surfactant concentration and oil type. Colloids and Surfaces A: Physicochemical and Engineering Aspects. 2013;**438**:148-158

[28] Koczo K, Lobo LA, Wasan DT. Effect of oil on foam stability: Aqueous foams stabilized by emulsions. Journal of Colloid and Interface Science. 1992;**150**(2):492-506

[29] Osei-Bonsu K, Shokri N, Grassia P. Foam stability in the presence and absence of hydrocarbons: From bubble- to bulk-scale. Colloids and Surfaces A: Physicochemical and Engineering Aspects. 2015;**481**:514-526

[30] Schramm LL, Novosad JJ. The destabilization of foams for improved

oil recovery by crude oils: Effect of the nature of the oil. Journal of Petroleum Science and Engineering. 1992;7(1):77-90

[31] Kapetas L, Vincent Bonnieu S, Danelis S, Rossen WR, Farajzadeh R, Eftekhari AA, et al. Effect of temperature on foam flow in porous media. Journal of Industrial and Engineering Chemistry. 2016;**36**:229-237

[32] Yekeen N, Manan MA, Idris AK, Padmanabhan E, Junin R, Samin AM, et al. A comprehensive review of experimental studies of nanoparticles-stabilized foam for enhanced oil recovery. Journal of Petroleum Science and Engineering. 2018;**164**:43-74

[33] Zhu J, Yang Z, Li X, Hou L, Xie S. Experimental study on the microscopic characteristics of foams stabilized by viscoelastic surfactant and nanoparticles. Colloids and Surfaces A: Physicochemical and Engineering Aspects. 2019;**572**:88-96

[34] Kim I, Taghavy A, DiCarlo D, Huh C. Aggregation of silica nanoparticles and its impact on particle mobility under high-salinity conditions. Journal of Petroleum Science and Engineering. 2015;**133**:376-383

[35] Ibrahim AF, Nasr-El-Din HA. Effect of water salinity on coal wettability during CO_2 sequestration in coal seams. Energy & Fuels. 2016;**30**(9):7532-7542

[36] Worthen AJ, Bagaria HG, Chen Y, Bryant SL, Huh C, Johnston KP. Nanoparticle-stabilized carbon dioxide-in-water foams with fine texture. Journal of Colloid and Interface Science. 2013;**391**:142-151

[37] Emrani AS, Nasr-El-Din HA. Stabilizing CO_2 foam by use of nanoparticles. SPE Journal. 2017;**22**(02):494-504

[38] Bayat AE, Rajaei K, Junin R. Assessing the effects of nanoparticle type and concentration on the stability of CO_2 foams and the performance in enhanced oil recovery. Colloids and Surfaces A: Physicochemical and Engineering Aspects. 2016;**511**:222-231

[39] Nasr-El-Din H, Chesson J, Cawiezel K, Devine C. Laboratory Investigation and Field Evaluation of Foamed VES Diversion Applied During Coiled-Tubing Matrix-Acid Treatment. paper SPE.99651:4-5

[40] Fang J, Dai C, Yan Y, Yan Z, You Q, editors. Enhanced foam stability by adding dispersed particle gel: A new 3-phase foam study. In: SPE Asia Pacific Enhanced Oil Recovery Conference. Society of Petroleum Engineers; 2015

[41] Wei P, Pu W, Sun L, Pu Y, Li D, Chen Y. Role of water-soluble polymer on foam-injection process for enhancing oil recovery. Journal of Industrial and Engineering Chemistry. 2018;**65**(01):280-289